SpringerBriefs in Astronomy

Series Editors

Martin A. Ratcliffe
Wolfgang Hillebrandt
Michael Inglis

For further volumes:
http://www.springer.com/series/10090

If the basic unit of geomorphology is the drainage basin . . .

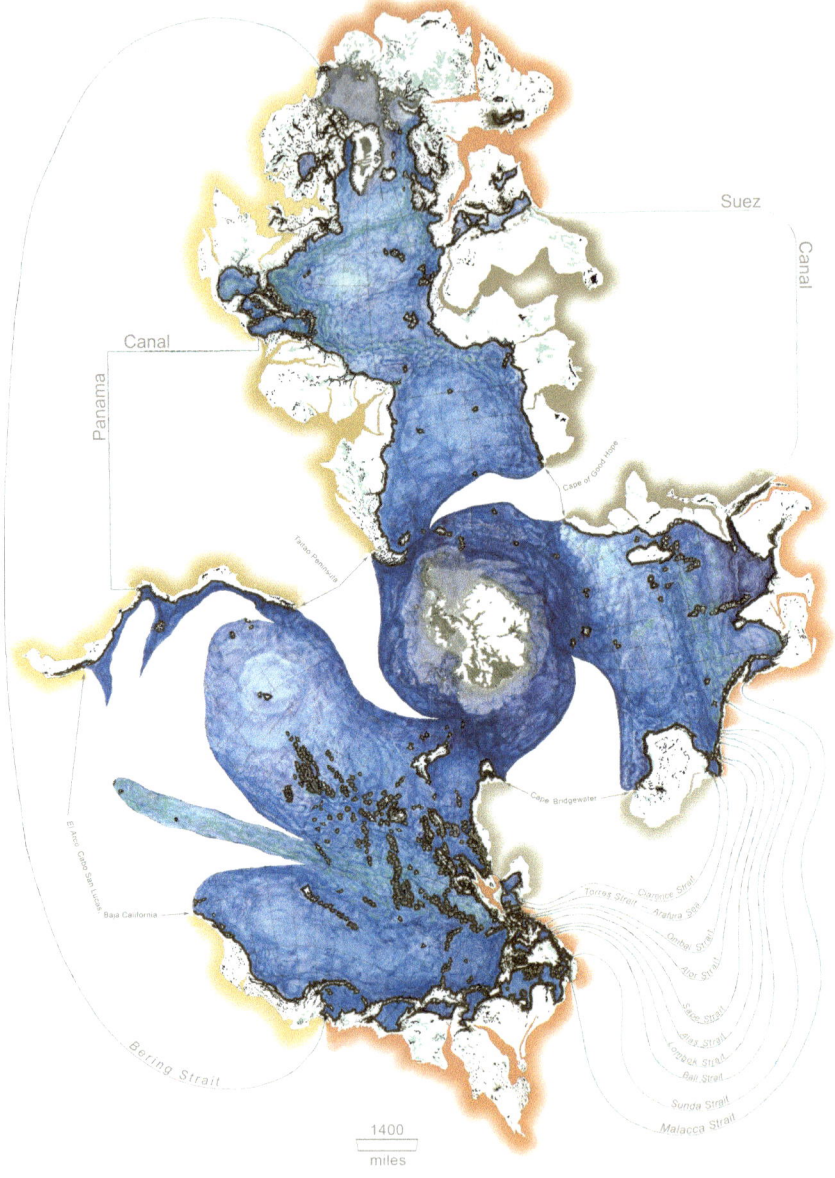

A Constant-Scale Natural Boundary Map of Earth *Showing Watersheds adjacent their respective Oceans (broken by Currents), Natural Lakes (full), Rivers (tidal length), Wetlands, Depressions & Ice.* 1996–2013 Chuck Clark, architect.

Pamela Elizabeth Clark • Chuck Clark

Constant-Scale Natural Boundary Mapping to Reveal Global and Cosmic Processes

 Springer

Pamela Elizabeth Clark
Institute of Astrophysics and
    Computational Sciences
Catholic University of America
Washington, DC, USA

Chuck Clark
Chuck Clark, Architect
Atlanta, GA, USA

ISSN 2191-9100   ISSN 2191-9119 (electronic)
ISBN 978-1-4614-7761-7   ISBN 978-1-4614-7762-4 (eBook)
DOI 10.1007/978-1-4614-7762-4
Springer New York Heidelberg Dordrecht London

Library of Congress Control Number: 2013943593

© The Author(s) 2013
This work is subject to copyright. All rights are reserved by the Publisher, whether the whole or part of
the material is concerned, specifically the rights of translation, reprinting, reuse of illustrations, recitation,
broadcasting, reproduction on microfilms or in any other physical way, and transmission or information
storage and retrieval, electronic adaptation, computer software, or by similar or dissimilar methodology
now known or hereafter developed. Exempted from this legal reservation are brief excerpts in connection
with reviews or scholarly analysis or material supplied specifically for the purpose of being entered and
executed on a computer system, for exclusive use by the purchaser of the work. Duplication of this
publication or parts thereof is permitted only under the provisions of the Copyright Law of the Publisher's
location, in its current version, and permission for use must always be obtained from Springer.
Permissions for use may be obtained through RightsLink at the Copyright Clearance Center. Violations
are liable to prosecution under the respective Copyright Law.
The use of general descriptive names, registered names, trademarks, service marks, etc. in this publication
does not imply, even in the absence of a specific statement, that such names are exempt from the relevant
protective laws and regulations and therefore free for general use.
While the advice and information in this book are believed to be true and accurate at the date of
publication, neither the authors nor the editors nor the publisher can accept any legal responsibility for
any errors or omissions that may be made. The publisher makes no warranty, express or implied, with
respect to the material contained herein.

Illustrations © Chuck Clark 2013

Printed on acid-free paper

Springer is part of Springer Science+Business Media (www.springer.com)

This book is written in memory of Dr. Paul Daniel Lowman, Jr., (1931–2011). From NASA's earliest days, Paul was an influential innovator and tireless worker in orbital remote sensing and photointerpretation applied first to the Earth and then to other planetary objects. He was a true friend and advocate. (Photo courtesy of NASA GSFC).

Dedication

This book is dedicated to Paul Lowman and to those whose work in diverse fields inspired our paradigm for physical process–based combined mapping and 3D modeling.

We are especially grateful for the ongoing encouragement to carry on "outside the box" from our colleagues Phil Stooke, René De Hon, and Michael Purucker.

For vital assistance and fruitful discussions, we are indebted to polymath Athelstan Spilhaus (1911–1998); geophysicist Dave McAdoo; planetary scientists Mark Robinson, Peter Thomas, Gunther Kletetschka, and Marc Berthoud; extraterrestrial cartographer Trent Hare; mathematicians Steven Krantz and Nick Grossman; anthropologists Alan Dorian and Glen Doran; geographer Jarke van Wijk; cosmologist Alan Kogut; oceanographer Emil Petruncio; molecular chemist James Harrison; cognitive psychologist Marilyn Berman; reference librarian Lloyd Busch; photographer Sara Adkins; editor John Caudle; image wizard Bill Mills; and the remarkable model maker Tayfun Öner.

For dozens of conference posters and hundreds of handouts, we thank the generosity of Imaging Technologies.

This project's genesis traces to poet Robert Lee Frost (1874–1963), who gave the problem of inventing a new way to make world maps to coauthor Chuck Clark as an apprentice task to accompany entrance into the guild of architecture; to mathematician Marston Morse (1892–1977), who, using the contours of his hand, taught young Chuck the topography of Morse theory; and to architect jim hagan of Atlanta, GA, who in 1990 awakened Chuck's dormant task with a request to study certain global-scale aspects of the Great Pyramid. As Morse expressed it in a 1958 Author's Note added to his essay "Mathematics and the Arts," quoting Johannes Kepler:

> The roads by which [we] arrive at [our] insights…seem to me as worthy of wonder as those matters themselves.

Contents

1 Constant-Scale Natural Boundary Mapping in Context 1
   1.1   The Value of a Geometry-Based Mapping Paradigm 1
   1.2   Relationship to Conventional Cartographic Techniques................... 2
   1.3   Deriving Boundaries: Maxwellian Hills and Dales 11
   1.4   The Use of Boundaries... 12
   1.5   Relationship to 3D Modeling Techniques and Topology................. 17
   1.6   Relationship to Perspective-Based and Anamorphic Drawing 22

2 Constant-Scale Natural Boundary Mapping Technique 27
   2.1   Identifying Critical Boundaries, Unzipping and Zipping 27
   2.2   Making Closed Shapes and Adjusting Proportions 29
   2.3   Adjusting Internal Scale.. 29
   2.4   Drawing the Grid and Creating a Map... 32
   2.5   Folding 36
   2.6   Waterlining... 36
   2.7   Demonstration... 39
   2.8   Summary of Implications for Global Mapping................................ 42

3 Interpretation of CSNB Maps.. 43
   3.1   Nature of Processes and Resulting Boundaries 43
   3.2   Externally Driven Processes 44
   3.3   Internally Driven Processes 45
   3.4   Making Comparisons 46

4 Mapping the Earth... 47
   4.1   Earth's Dynamic Context 47
   4.2   Tectonic Activity... 48
   4.3   Watersheds, Watercourses, and Weather.. 51
   4.4   Ocean Currents.. 56

5 CSNB Mapping Applied to Regular Bodies.. 59
   5.1   Overview of Application... 59
   5.2   The Moon... 60
   5.3   Venus... 60
   5.4   Mars 63
   5.5   Future Applications.. 68

6 CSNB Mapping Applied to Irregular Bodies 71
   6.1   Overview of Application... 71
   6.2   Distribution of Features on Asteroid 433 Eros 72
   6.3   Comparison of Eros, Phobos, Deimos, and Ida 76
   6.4   Exploring Asteroid 25143 Itokawa 81
   6.5   Other Irregular Objects 85

7 Mapping the Sky 89
   7.1   Beyond Human Sight.. 89
   7.2   Cosmic Microwave Background....................................... 89
   7.3   CSNB Maps of CMB Anisotropy 92

8 The Future of CSNB Mapping... 95
   8.1   Status, Goals, and Motivation 95
   8.2   Methodology and Plan for 2D Mapping 96
   8.3   Methodology and Plan for 3D Modeling 96

References 101

Provenances 107

Acronyms 113

About the Authors.. 115

Chapter 1
Constant-Scale Natural Boundary Mapping in Context

1.1 The Value of a Geometry-Based Mapping Paradigm

Constant-Scale Natural Boundary (CSNB) mapping is a revolutionary approach to visualization that produces maps markedly different from, and yet complementary to, those produced by conventional 2D cartographic and 3D modeling techniques (Clark 2002, 2003, 2004a, b, 2005, 2007, 2011; Clark and Clark 2005, 2006a, b, 2007, 2009, 2010; Clark et al. 2006, 2007). Whereas conventional maps are built on predetermined grids or formulae (Spilhaus and Snyder 1991) which often distort the shapes and scale of prominent features that result from underlying processes, CSNB maps begin with well-defined boundaries, often found at the 'edges' of conventional projections, that result from natural processes. These boundaries are obvious regional-scale highs and lows in the parameter space of interest: elevation highs (ridges) or lows (troughs or valleys) for topography (geography) or cranial mapping (anthropology), or, analogously, in pressure or temperature (meteorology); surface roughness, texture, or scattering (photointerpretation); density of objects of interest or intensity of any defined signal for a range of disciplines. By directly illustrating spatial and dynamic relationships, CSNB maps provide greater insight into formation processes, revealing global-scale relationships and patterns in the distribution of analogous features, such as watersheds, mountain chains, prevailing currents and winds, explosions, or galactic clusters. This technique is particularly applicable to irregular bodies or systems with roughness on the scale of the body itself, such as asteroids or crania. In those cases, imposing a conformal grid distorts features beyond recognition. Finally, CSNB bridges the gulf between the 2D mapping and 3D modeling worlds, which typically require completely unrelated algorithms and assumptions, by allowing visualization of an object in either 2D or 3D with the same set of assumptions, and thereby facilitating the transition between the two.

The CSNB methodology requires geometric reasoning, with its emphasis on visualization, spatial awareness, and deductive logic. The historical weakness in deductive reasoning, uncertainty in the premise, turns out to be a strength in

P.E. Clark and C. Clark, *Constant-Scale Natural Boundary Mapping*
to Reveal Global and Cosmic Processes, SpringerBriefs in Astronomy,
DOI 10.1007/978-1-4614-7762-4_1, © The Author(s) 2013

map development and interpretation, which typically require multiple working hypotheses. In fact, the deductive process provides complementarity to the inductive reasoning approach used in the scientific method, which may be limited by too narrow a focus and thus lack of representative examples. A set of CSNB maps, each associated with a different set of assumptions for feature capture based on a different testable hypothesis for origin, can be generated, and their implications can be explored and evaluated on the basis of new or existing observations.

1.2   Relationship to Conventional Cartographic Techniques

To make maps with constant-scale natural boundaries is to make maps differently than we've been doing for the past 450 years (Clark and Clark 2006a, b), when mapmaking as we have come to know it, namely cartography, developed as a method for projecting a regular 3D object, smooth on the scale of its size, onto a 2D surface with a systematic grid system (longitude and latitude).

The concept of the projection of a sphere from a defined central point, typically the north pole, onto a plane was first developed by the Egyptians, and described by Ptolemy (Sidoli and Berggren 2007) as a planisphere, or stereographic projection (Fig. 1.1). The projection did not use a grid (latitude/longitude) system. However, such a projection caused major distortions in distances at the periphery of the map, and thus was not particularly useful for travel.

Boyer (1968) indicates that "In the sixteenth century, geographical explorations had widened horizons and created a need for better maps . . . new discoveries had outmoded medieval and ancient maps. In 1569, Mercator published the first map drawn up on the new principle (of perspective), introducing the map that bears his name and that approach, with later improvement, has been basic in cartography ever since." The first step in a Mercator projection is to think of a spherical earth inscribed within a right circular cylinder touching the earth along the equator (or some other great circle), and to project, from the center of the earth, points from the surface of the earth onto the cylinder. If the cylinder is then cut along an element and flattened out, the meridians and parallels on the earth will have been transformed into a rectangular network of lines (Fig. 1.1). Edward Wright (Boyer 1968) developed the theoretical basis of the cartography, as well as the Mercator projection in 1599, by computing the functional relationship between map distance from the equator and degree of latitude.

Dürer (Panofsky 1943) proposed a novel 'prototopological' approach for representing the surfaces of solids as a coherent net of adjoining naturally bounded facets that could be folded or cut to represent that surface in three dimensions (Fig. 1.1). This idea challenged the limitations and utility of the strictly two-dimensional representations. Panofsky's invention of the term *prototopological* was not capricious: he was one of the founding faculty of the Institute for Advanced Study in 1935, and a close colleague of Institute mathematician Marston Morse, one of the framers of modern topology. Morse, by extending the work of Maxwell (1870), established an eponymous branch of mathematics.

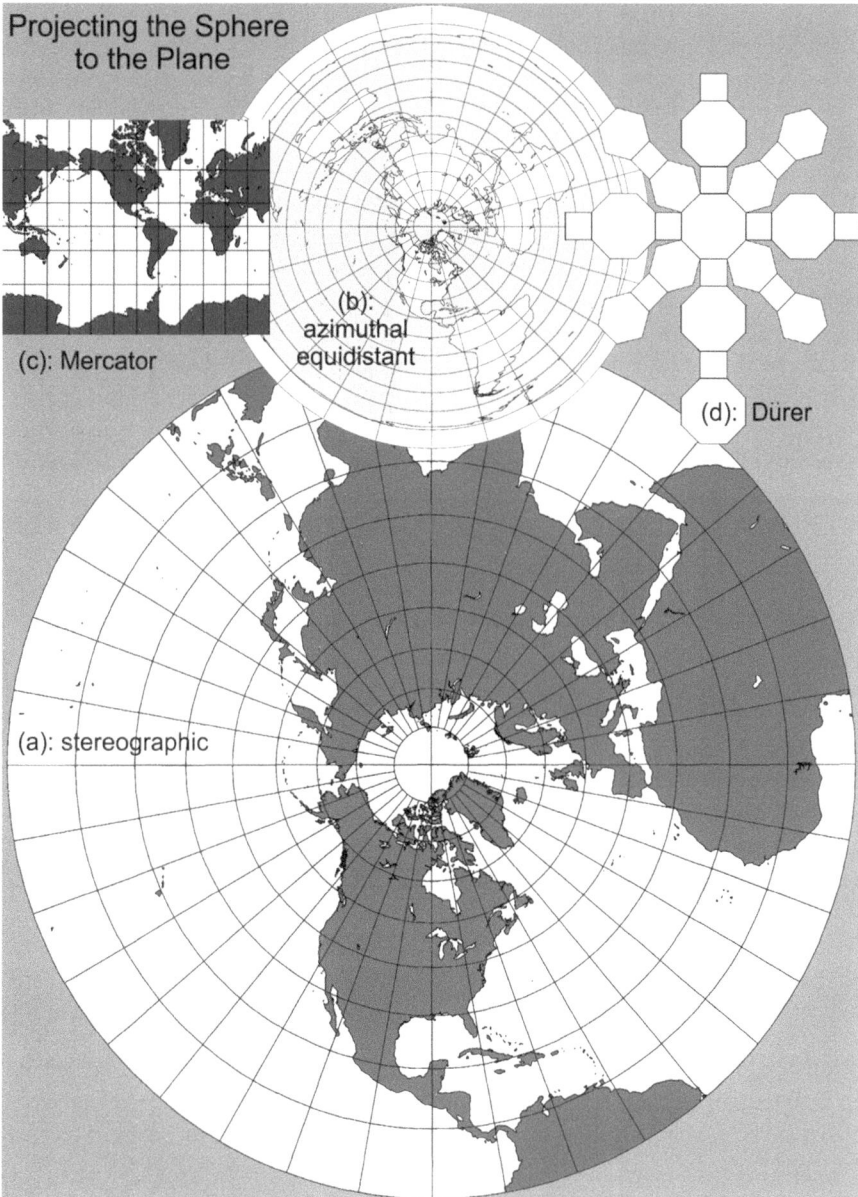

Fig. 1.1 Projecting a sphere onto a plane (Clark 2003) using (**a**) stereographic, (**b**) azimuthal equidistant, (**c**) Mercator, and (**d**) Dürer's projections (Clark 2003) as discussed in the text. (**a**), (**b**), and (**c**) courtesy of USGS (Snyder and Voxland 1989)

If we replace Dürer's constant azimuth between vertices with CSNB's natural (irregular) azimuth, Panofsky's prototopological quality still applies: the map still folds. Although Dürer didn't construct any maps from his nets, Dürerian nets have become a standard reference in cartography (e.g., Fisher and Miller 1944; Snyder 1993).

Understanding scale, and where constant-scale occurs in a given projection, is critical for map interpretation (Fig. 1.2). Snyder (1977, 1981) and others called for maps with constant scale along the ground track of the LandSat 7 satellite. Curiously, Snyder's last collaborations included global maps with natural boundaries (Snyder 1993), involving attempts to separate and illustrate the continuity in ocean and land regions with complicated cartographic projections (Fig. 1.3). Their innovation was to cut the inscribing solid along an irregular (natural) line instead of an element (straight line), leaving unchanged what has remained unchanged since Mercator: an inscribing solid and its attendant algebraic formulation begins the process. They apparently did not notice that by combining constant-scale (crucial for control of distortion) with natural boundary (crucial for articulate global context), they would be creating the basis for a new, topologically informed class of maps with constant-scale natural boundaries (CSNB).

Recently, van Wijk (2008) has pioneered a digital route to Spilhaus-type global maps, and Stooke (1998) has shown that novel methods can produce useful maps of objects, such as asteroids, poorly suited to traditional cartography (Fig. 1.4). However, his methods are uniquely tailored for each body, and thus not generalizable for an over-arching system, as CSNB appears to be. The equal-area map method of Berthoud (2005) was developed to illustrate surface-feature density on irregular bodies. The method provides coverage of 433 Eros (Fig. 1.4) in two fixed maps. However, it generates overlaps when applied to more irregular asteroids such as Kleopatra. CSNB can map Eros into one adjustable, and reconfigurable image, varying from highly compact to highly segmented (Clark and Clark 2006a) (Fig. 1.5). Please refer to Chap. 6 for a more complete discussion of this subject.

Nevertheless, cartography has become the prevailing mapping paradigm, despite attempts to use topology-based projections derived from natural features. Other innovations (Greeley and Batson 1990) have included rapid digitization and the development of the Global Information System, allowing overlay and multi-dimensional processing of mappable parameters, proliferation of new projections for a wider range of applications, development of non-visual sub-disciplines such as analytical cartography, and production of map novelties such as the world projected onto a Möbius strip and a torus (Tobler 1961; Snyder and Voxland 1989).

The digitization process was a crucial innovation, making possible the exploration of space via the collection, transmission, and production of maps from staggering amounts of data from Earth orbit and beyond (Lowman 2002). Further, computerization of stereograph and animation made 3D reconstructed fly-over animations possible with progressively less effort.

However, these new techniques, even when combined, left the prevailing perspective-based mapping paradigm fundamentally unaltered, and focused on details as opposed to global context as the effective resolution of measurements improved. Such points of view inevitably focus on small areas where distortion is

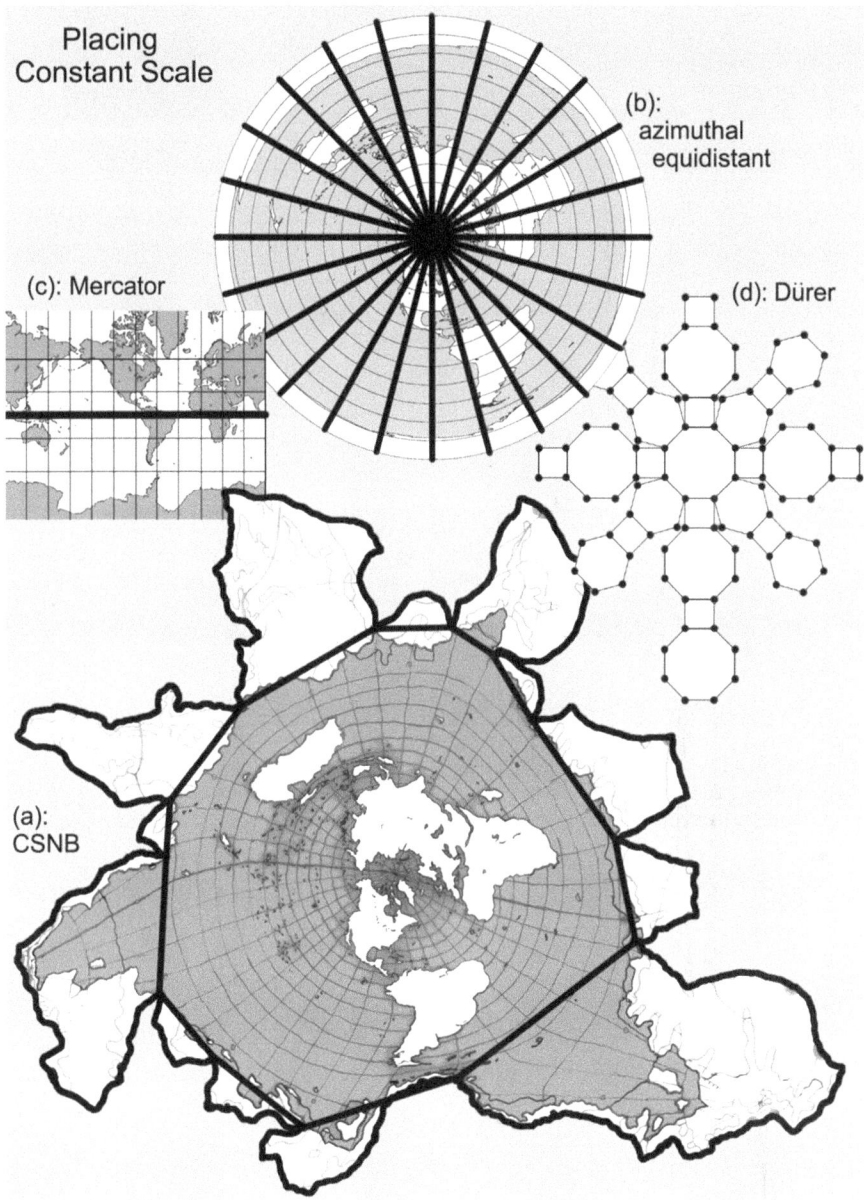

Fig. 1.2 Location of constant scale (heavy *lines*) in four projections including CSNB (Clark 2003). (**a**) Source data courtesy of NASA GSFC. (**b**) and (**c**) courtesy of USGS (Snyder and Voxland 1989)

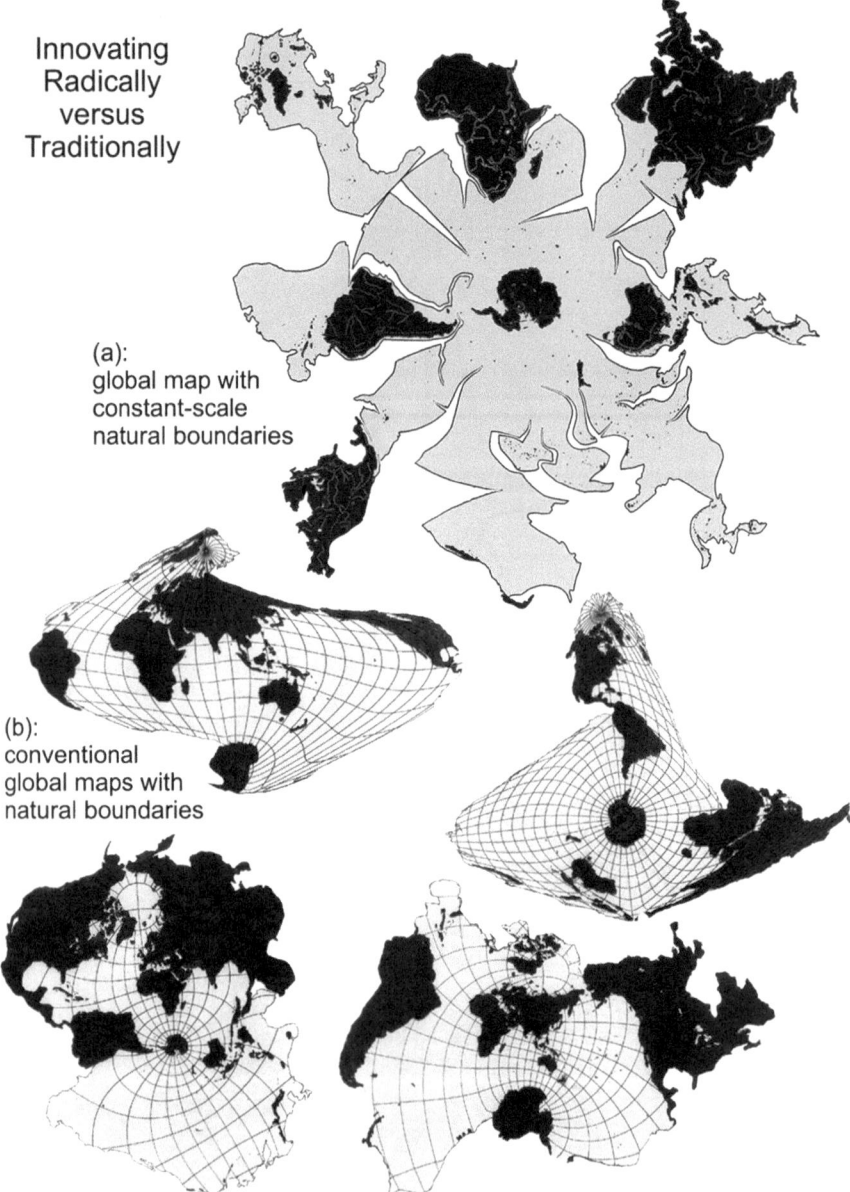

Innovating
Radically
versus
Traditionally

(a):
global map with
constant-scale
natural boundaries

(b):
conventional
global maps with
natural boundaries

Fig. 1.3 Separating land from ocean via (**a**) CSNB undistorted boundary method and (**b**) conventional method (Clark 2007). (**a**) Source data Spilhaus (1991). (**b**) Spilhaus (1991) (Courtesy of American Philosophical Society)

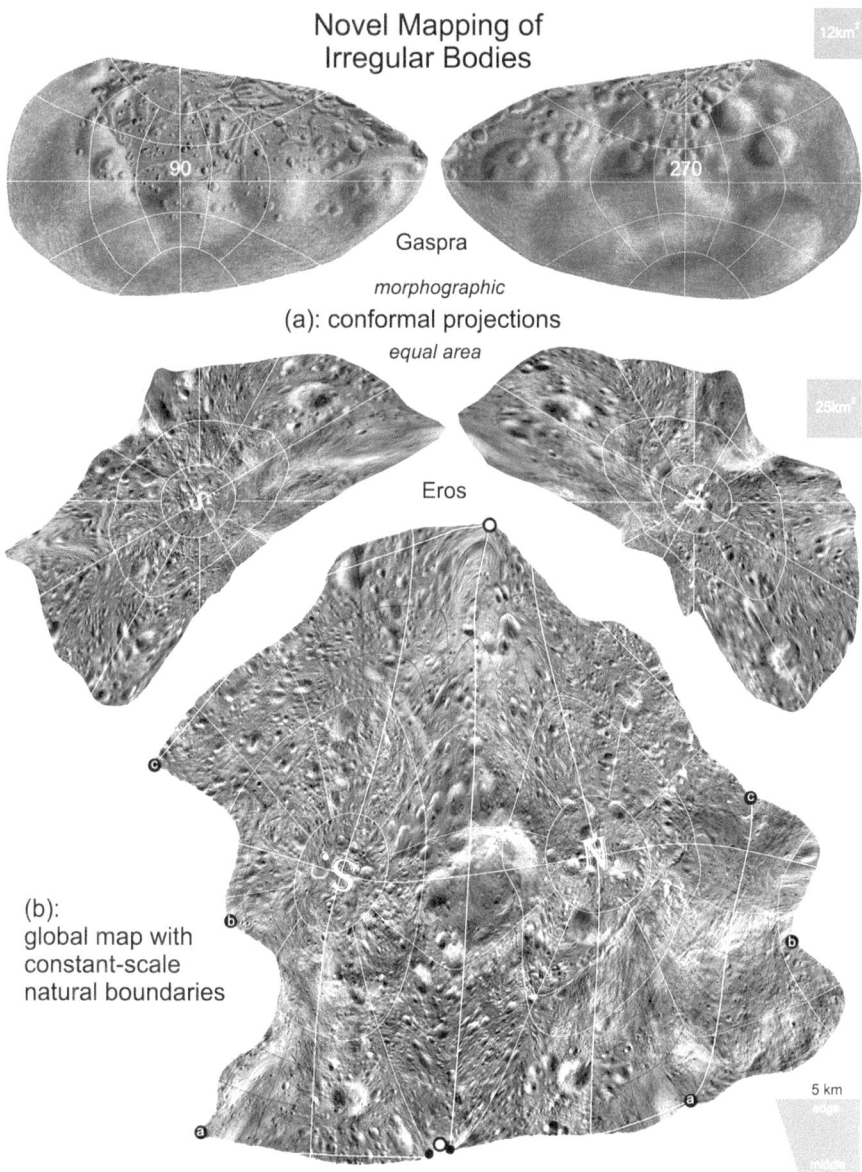

Fig. 1.4 Novel approaches for mapping irregular bodies applied to asteroid 433 Eros. (**a**) (*top*) Courtesy of P. Stooke. (*middle*) Projection courtesy of (Berthoud 2005) and source data of P. Stooke Eros photomosaic. (**b**) Source data P. Stooke Eros photomosaic

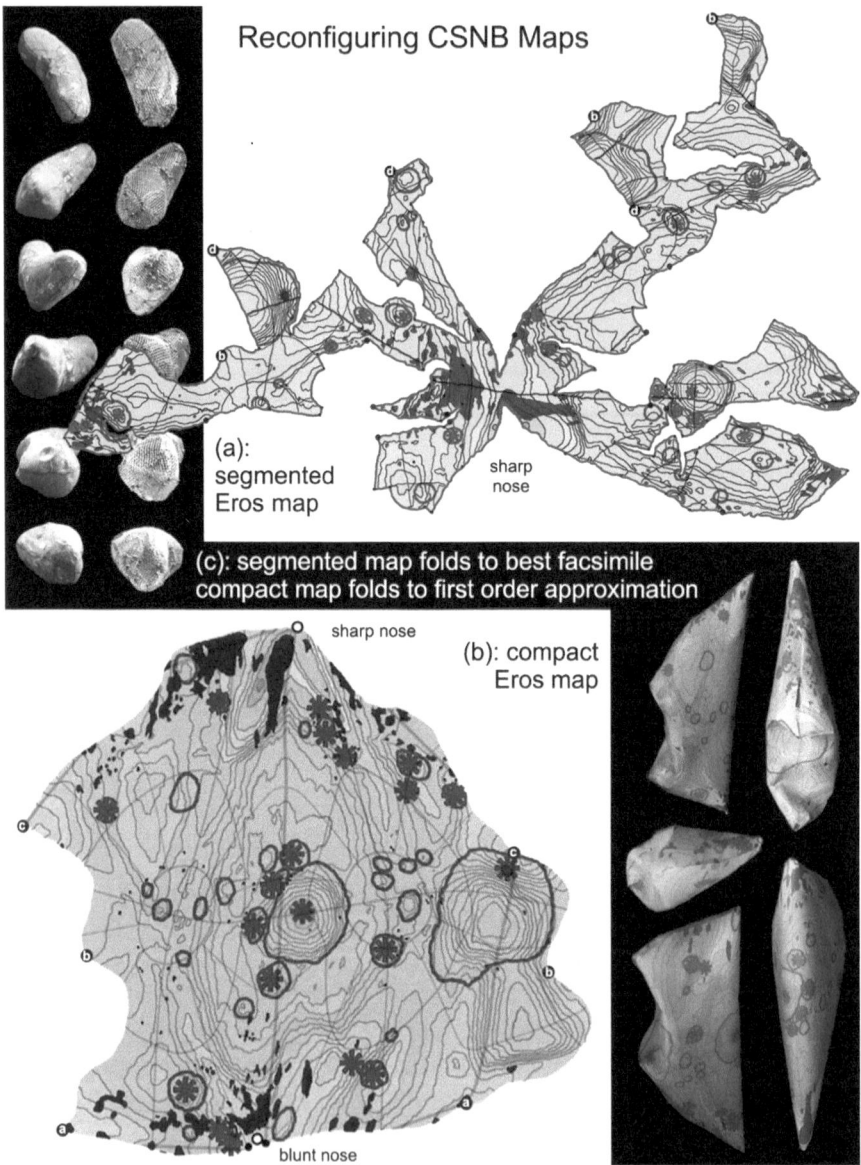

Fig. 1.5 Examples of CSNB maps and models for irregular objects (Clark and Clark 2006a, b). (**a**) and (**b**) Source data Cheng et al. 2002 (topography), Mantz et al. 2002 (ponds). (**c**) Source data (3D models) Peter Thomas

Table 1.1 Comparison of conventional projections for regular bodies

Type	Advantages	Disadvantages	Examples
Conformal	Preserves local shape, angular relationships, and scale over small areas	Distortions in size and distance as function of latitude	Mercator, simple cylindrical
Azimuthal	Preserves direction from line or point on global scale	Distortion in shape as a function of distance from line or point	Polar, Albers
Equal-Area	Preserves area on global scale	Non-conformal, distortions in shape	Mollweide, Lambert azimuthal equal-area
Equidistant	Preserves distance on global scale	Non-conformal, distortions in shape	Azimuthal equidistant, Werner, sinusoidal

least, and thus can harbor inconsistencies when used to illustrate relationships between analogous features on a larger scale. This is illustrated by the current difficulty in understanding the stratigraphic record of Sinus Meridiani on Mars (Edgett 2005). Because ground truth available in the form of in situ measurements from apparently related sites is limited, a more comprehensive global perspective is needed.

Modern maps are still largely cartographic applications, including global or local topographic maps, thematic maps, and selected parameter (e.g., cloud cover or vegetation cover, typically through signatures derived from visible, infrared, or microwave bands) maps displaying specialized information for selected audiences, as well as local orienteering maps containing accurate topography, landmarks, and features that indicate landscape mobility degree of difficulty. The availability of GPS has revolutionized the production of cartographically-based field maps from measurements made and landmarks noted in the field in real time. A smaller number of specialized maps are topology-based, showing the 'shape' of a system and relationships between elements within it, such as subway maps.

Projections typically assume regular, spherical objects, although the regular celestial objects are typically oblate spheroids, and most solar system bodies, the asteroids, are completely irregular objects. However, any three-dimensional object can be mapped, because projections flatten any continuous surface enclosing a three-dimensional object into a plane with a two-dimensional grid system (latitude and longitude). Map projection developable surfaces for unrolling into a sheet may be a cylinder around an axis (e.g., Mercator), a cone around a point and an axis (e.g., Albers), or a plane around a point (e.g., azimuthal polar projections). The aspect (relationship to the axis of rotation) may be parallel (normal) or transverse (perpendicular) to the axis or rotation, or somewhere between the two (oblique). The resulting projection introduces progressively greater distortion to a greater and greater extent as the object becomes less regular, the distance from the central point or line greater, or the projection less appropriate to the application. Distortion may be introduced in area, shape, direction, distance, or scale as a function of latitude and longitude. Projections may preserve local shape (conformal), direction from a line or point (azimuthal), area (equal-area), or distance (equidistant). Projections are chosen to minimize distortion in the parameters most important for that application, as illustrated in Table 1.1. For example, local maps typically employ transverse Mercator to

Table 1.2 Comparison of CSNB and conventional mapping techniques

Feature	Conventional capability	CSNB advantages	Mission benefit
Local Minimize distortion, maintain proper scale	Distortion depends on relationship to 'center'. Cannot maintain constant resolution	Constant undistorted boundary scale represents underlying processes at appropriate resolution, little to no distortion	Decrease resources consumed, including cost, for productive science, as target selection or route planning, during or post-mission either with ground or onboard processing capability, yielding flexible, agile, resilient map/ model process for multiple domains or surveys
Global Facilitate Interpretation of origins, processes	Maps start with arbitrary grid, distort proportions, weaken global interpretation	Map starts with naturally formed ridges and troughs in parameter space that shape the landscape	
Visual Utility for 2D and 3D	Choose one or the other, not both	Identify boundaries from 'cloud of points'; segment and project in 2D; reconnect and project in 3D	
Mission Automate mapping, modeling process	Current 3D techniques require highest resolution, are time-intensive, and difficult to automate	Produce CSNB maps of growing complexity with growing resolution as a rotating object is approached	

CSNB Domains: astrophysics, meteorology, geophysics, planetology, heliophysics, geography, oceanography, anthropology, physical chemistry

preserve angular relationships (conformality) and scale over small areas. For global maps, if equal-area projections are used to preserve relative size of continents regardless of latitude (e.g., Mollweide, Lambert) or equidistant projections to preserve distances along grid lines or between points (plate carrée, azimuthal equidistant, Werner, sinusoidal) then shapes may be distorted (Fig. 1.1). Thus, compromise projections may be used instead, such as the Winkel tripel favored by the National Geographic magazine. Distortions increase with distance from the projection-defined map 'center' (a point or line network of constant scale) (Fig. 1.2). CSNB maps maintain least distortion not at the centers, which are not necessarily very interesting, but instead along the boundaries that constitute the map edges and reflect the most significant physical processes. The relationships between scale, distortion, and conformality for CSNB and cartographic projections are summarized in Table 1.2.

Modern flattening transform procedures have evolved from algebraic considerations of projection geometry and developable surfaces. Modern projections no longer use perspective; instead they employ mathematical functions to transform coordinates from a curving surface to a flat one. However, even in our topologically conscious and digital capable era, the rules of perspective continue to underlie those functions (Boyer 1968). Perspective devices from the Renaissance include the picture plane, horizon, station point, vanishing point, and—vital for this exposition—the *plane trace*: a line containing the vanishing points of all systems of parallels in a picture in linear perspective. A plane trace is functionally similar, at least from a geometric point of view, to the edge of a CSNB map.

The mathematics of perspective underlies not only the capability to fashion the illusion of three-dimensional space onto a two-dimensional surface, but also the inverse capability to transform the surface of a three-dimensional object such as a planet onto a two-dimensional surface, i.e., to make maps of objects that are enclosed systems. For the cartographer, algebraic shortcuts have long been available. For the geometer, both capabilities have historically involved the tedious plotting of points. The introduction of image processing paradigms to identify natural boundaries on regular objects or facet-edges of irregular ones should considerably reduce labor and increase speed of producing CSNB maps. This topic will be discussed in Chap. 8.

A *plane trace* is a line in perspective drawn through a vanishing point at an angle to the horizon, and is the sum of vanishing points of all similarly sloped lines in pictorial space, wherever they are in the picture plane. In one sense, this line should be infinitely thick, but in the theory of perspective, it exists at an infinite distance, and therefore appears as a line at a designated slope. When used in tandem with a second plane trace, their shared point is the vanishing point of lines lying in both planes, i.e. the planes' intersection. Intersecting two plane traces is a common operation in advanced perspective construction; it locates the point where lines shared by two planes vanish, when both planes are skew to both the picture plane and the ground plane, e.g., the intersection of a dormer roof with the main roof. In a CSNB map, poles are located by finding the point where three such traces meet.

1.3   Deriving Boundaries: Maxwellian Hills and Dales

The CSNB approach is linked conceptually to James Clerk Maxwell's seminal paper "On Hills and Dales" (1870). Maxwell saw the Earth's surface as variations from an equilibrious surface, in terms of potential work. The surface could be specified, as sea or other user-defined level. This 'watermark' can be visualized as a contour line on a topography map. If that contour were the elevation of the lowest low (a pit), the entire surface becomes a *hill*, if that contour were at the highest high (a peak), the entire surface would become a *dale*.

Perhaps this is obvious, but more significant from our standpoint are local maxima or minima in hill or dale complexes that form the basis of regionally extensive networks of hills, i.e., ridges, and networks of dales, i.e., valleys or troughs. Features that constrain hills and dales are known as saddles, passes, or bars. To visualize, think of a water level starting above the highest extremity and dropping gradually to uncover the highest peaks until a pass, a connection, is uncovered between two peaks. Positive slopes in the direction of the peaks and negative slopes in other directions surround such a saddle. A line between the summits along relatively shallow slopes (compared to vertical drops in roughly perpendicular direction) through the saddle would constitute the ridge. A ridge is recognizable as a string of positive contour rings separated by saddles. Similarly, the water level could start at the lowest extremity and fill pits until a bar separating two pits is about to be covered. Analogously to the pass, negative slopes in the direction of the pits and positive slopes in other directions surround the bar. A line between the pits through the saddle constitutes the trough or valley. A trough is recognizable as a string of negative contour rings separated by saddles. Networks of hills are watersheds, and ridges are 'lines of watershed'. Networks of dales are water sinks, and troughs are 'lines of watercourse'. See Fig. 1.6 for Maxwell's 1870 chart. Lines of watercourse surround hills, and lines of watershed surround dales. Ridges terminate in summits and troughs terminate in pits, and form complementary branching networks known as *trees* (Maizlish 2003), which intersect at common points called *saddles*. Note that both passes and bars are saddles, and may be the same feature, renamed as context shifts from top-down, when identifying ridges, to bottom-up, when identifying troughs.

CSNB maps are composed by the same concepts behind Morse Theory, a calculus of variations in the large, which "keeps the books" on maxima and minima, hills and dales, and seeks saddles of criticality. As Morse said, "often in physical processes it is not a peak or a pit which triggers sudden change, but rather a pass." Because complementary data structures share common saddles, complex maps are possible, blending districts, here using a portion of ridge, there a portion of valley, an approach which could be essential in identifying plate boundaries. Typically, (e.g., hydrological maps) global-scale ridges and troughs are intimately associated with distinct regions, and we generate complementary ridge and trough maps.

1.4   The Use of Boundaries

The CSNB approach revolutionizes how surfaces are visualized. Conventionally, mappers ask, "What is the study area?" and then select maps that place this at the middle. If global perspective is required because global-scale issues are involved, the conventional approach is to refine the choice of map projection to achieve least-worst peripheral distortion. The CSNB user asks, "What boundaries are critical?" This directly establishes the map.

Construction of CSNB maps is completely different from orthodox cartography. Conventional cartographers use algebraic techniques as a necessary shortcut to the tedious plotting of points, and employ the longitude-latitude grid system in a routine manner. The "improvements" which Boyer cites, e.g., the simple cylindrical versus the Mercator, have been efforts to reduce edge stretching; as we would put it, to approach constant-scale edge as a limiting ideal. CSNB begins with non-distorted constant-scale edges, and then proceeds backwards (inwards) to reestablish the grid

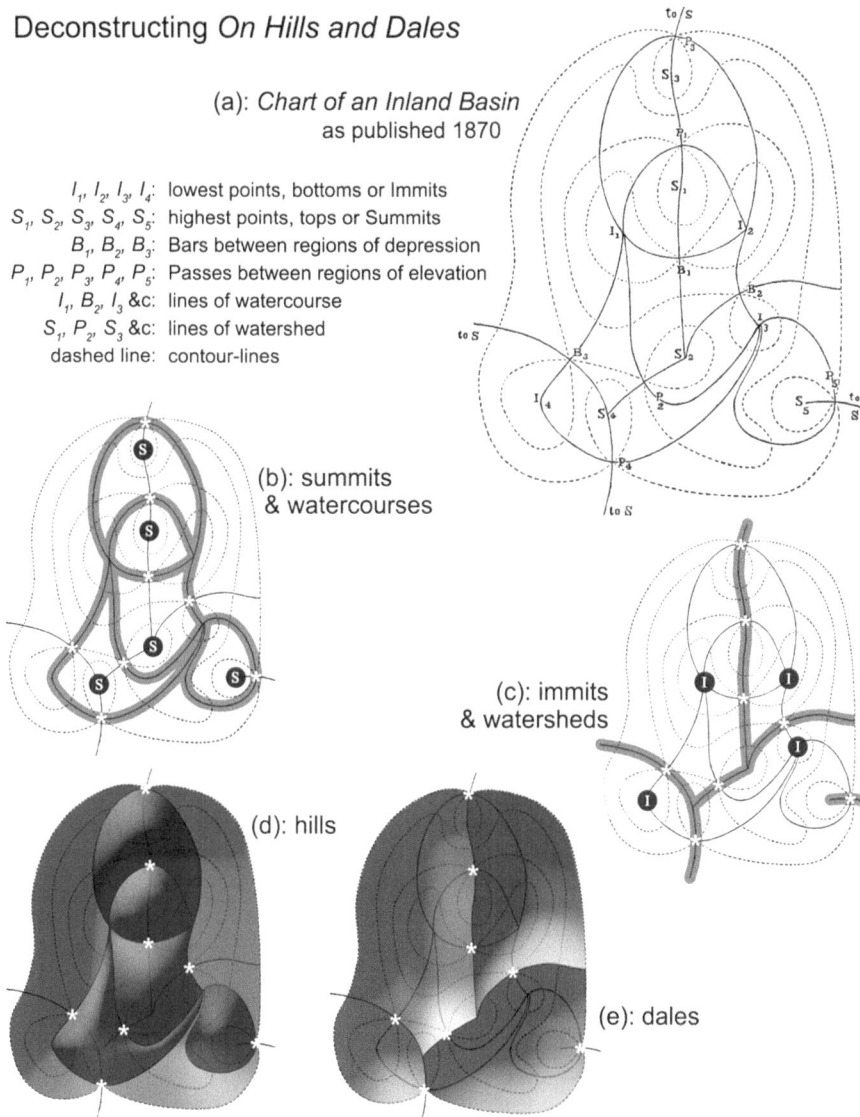

Deconstructing *On Hills and Dales*

(a): *Chart of an Inland Basin*
as published 1870

I_1, I_2, I_3, I_4:   lowest points, bottoms or Immits
S_1, S_2, S_3, S_4, S_5:   highest points, tops or Summits
B_1, B_2, B_3:   Bars between regions of depression
P_1, P_2, P_3, P_4, P_5:   Passes between regions of elevation
I_1, B_2, I_3 &c: lines of watercourse
S_1, P_2, S_3 &c: lines of watershed
dashed line: contour-lines

(b): summits
& watercourses

(c): immits
& watersheds

(d): hills

(e): dales

Fig. 1.6 (**a**) Maxwell (1870) Hills and Dales. Hills focus on summits, *dotted* in (**b**), and dales focus on 'immits' or *bottoms*, *dotted* in (**c**). *Lines* of watercourse and watershed intersect at *bars* and passes. Thus a body's entire surface may be divided into hills (**d**) or dales (**d**), each point on the surface belonging to a certain dale and also to a certain hill. (**a**) C.S. Clark modified Maxwell (1870) Chart of Inland Basin, Courtesy Taylor and Francis, Ltd., Philosophical Magazine; (**b–e**) (Clark 2003)

system that organizes the field of the map. The principles of contiguity and a smoothly varying scale between neighborhoods are our guide. CSNB allows proportionate (conformal) maps to be generated that are composed by, and only by, relevant critical boundaries. When constant scale is rigorously located at the map periphery, a shift in perception arises that reveals new patterns. In this light the azimuthal equidistant projection, which interrupts the surface at a single point, is a special case of CSNB, where the natural boundary is a point with physical meaning, as a pole, mega-quake, or, say, the impact responsible for the Moon's South Pole–Aitken Basin.

The emphasis on and control of boundaries shifts scientists' interest and attention away from the middle of the map and toward meaningful information distributed throughout the entire map. The greater the ratio of territory associated with critical boundaries to territory not associated with boundaries, the greater the map fidelity to the shape of the object of interest. The map itself becomes a unique visual symbol, and direct geometric expression of geomorphological variability of the surface. This is true whether the 'topographic' variable is elevation, or another parameter such as electron density of a molecule. In this way, the surface is an expression of positive or negative potential from an established mean.

Perspective is not used for the basic transformation. CSNB shuns projective geometry ('sheet goods' wrapping a globe) in favor of a one-dimensional tree, open-ended but otherwise analogous to a surveyor's metes and bounds, which we then transform, using radial unbending, into a series of planar surfaces separated by boundaries in a two-dimensional sheet (Fig. 1.6). Metes and bounds was the original English system of land surveying, based on walking and describing the encompassing boundary in terms of a series of traces, each with associated lengths and compass direction (metes), and bounds (e.g., streams or ridge lines) with benchmarks where two traces met. The tree contains the map's metrical 'benchmarks.' The flattened tree is a mathematically interesting interim state because the object's surface may be imagined as existing on the plane, extending to infinity, much like Mercator's map stretches from the equator to infinitely distant poles. With a CSNB tree, the plane's infinite and unbounded surface substitutes for the object's finite and bounded surface, as further discussed in Sects. 2.2 and 2.7.

The CSNB method restores the transverse dimension via bifurcation and chain-like manipulations of the tree (Fig. 1.7). Overlaps have not been problematic, even at concavities, but do generate ambiguity, which could complicate coding. Cross-map ratios are preserved, using a method discussed in detail in Chap. 2.

Higher resolution refinements can be easily introduced by introducing progressively more branching to the existing 'trees'. Higher resolution refinements, though available, do not have to be added, and in fact may obscure the larger pattern. The boundary (and thus the eventual map) may be manipulated while at the tree-stage, may be pruned or extended, as new features or patterns become more obvious, altering the eventual map in ways that depend on inductive reasoning rather than mathematical models (Fig. 1.8).

CSNB maps transform to three-dimensional objects. Without recourse to an arbitrary regular three-dimensional form or surface-filling shapes such as triangular facets, shortcuts typically used in 3D modeling, CSNB maps capture a body's natural

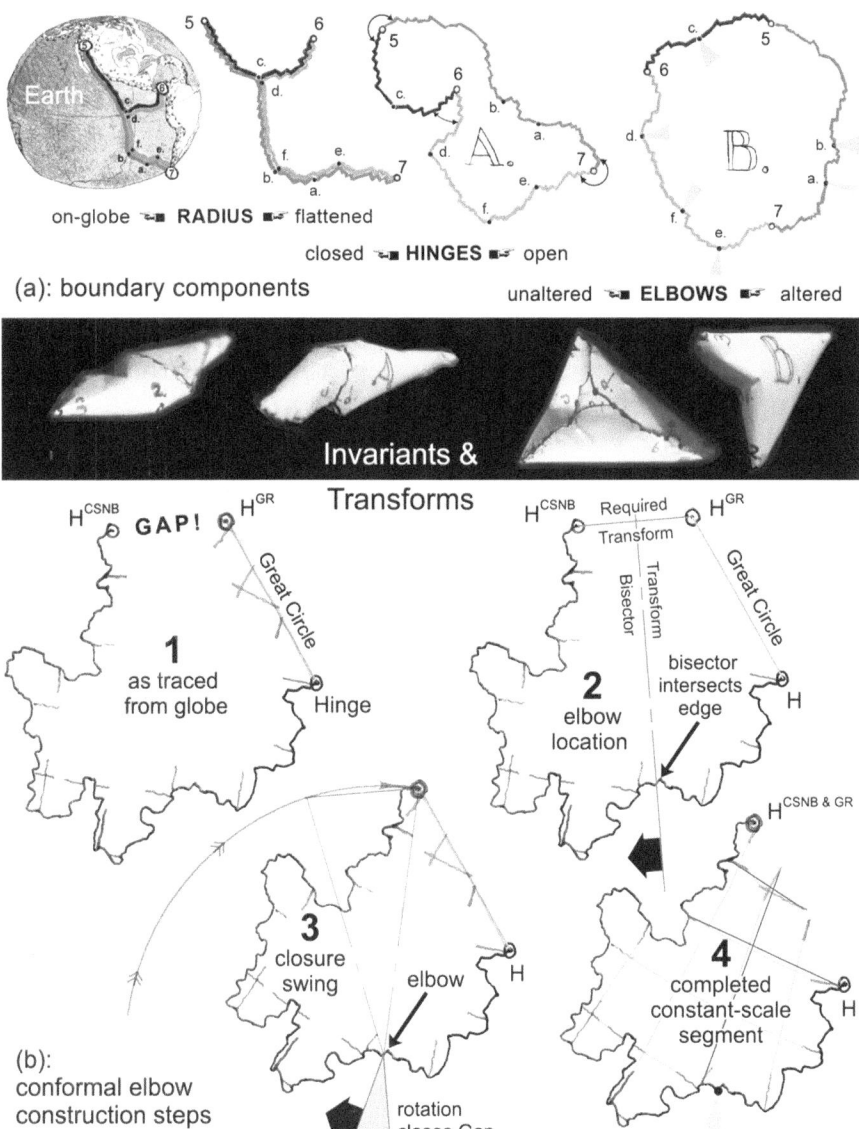

on-globe ⊶■ **RADIUS** ■⊶ flattened

closed ⊶■ **HINGES** ■⊶ open

(a): boundary components

unaltered ⊶■ **ELBOWS** ■⊶ altered

Invariants & Transforms

(b): conformal elbow construction steps

Fig. 1.7 Boundaries as feature edges treated as metes and bounds, transformed and reconnected using hinges and elbows as described in the text

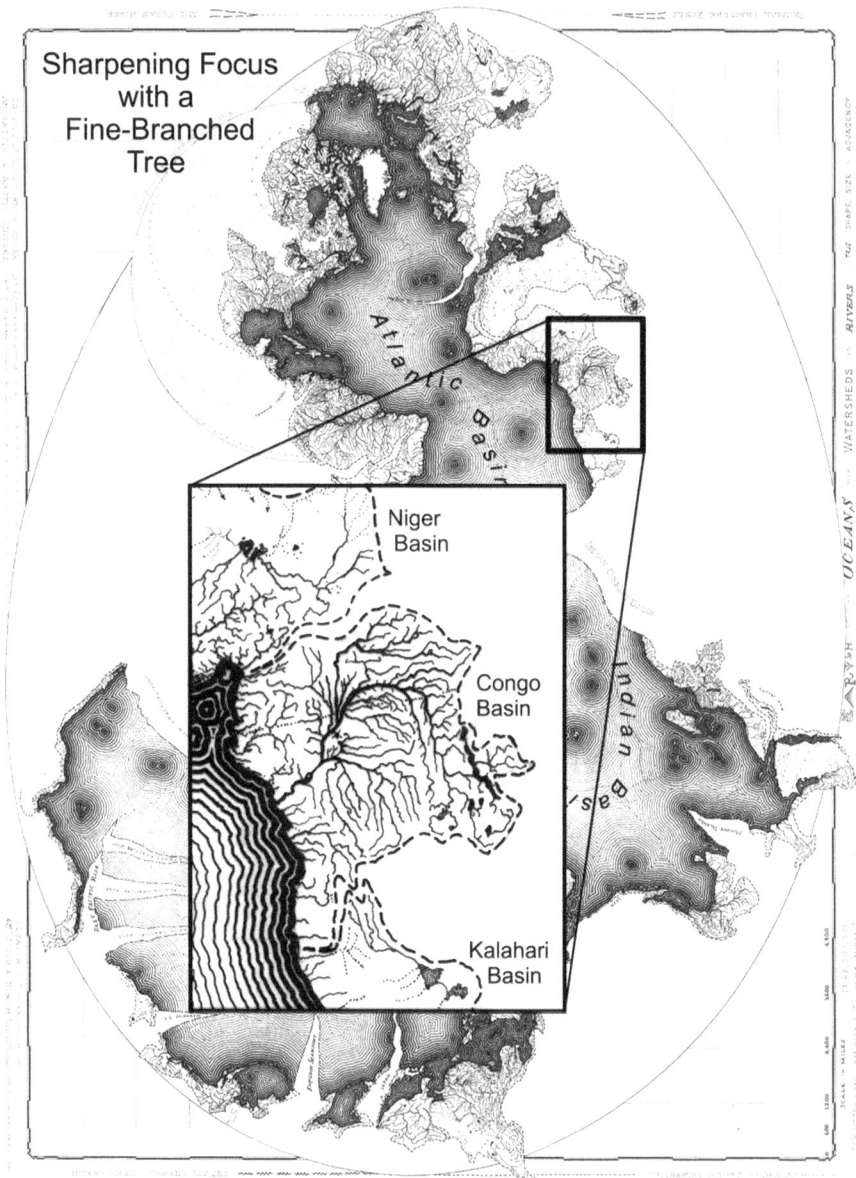

Fig. 1.8 Growth of trees to provide progressively greater detail as described in the text. Created by Chuck Clark in 1994

prototopology, to use Panofsky's 1943 term. Highly interrupted CSNB maps fold to excellent facsimiles, and may also model internal structure, such as continental roots (Fig. 1.9). Compact CSNB maps fold to condensations or first-order approximations of the object (Fig. 1.5). CSNB method restores the transverse dimension via bifurcation and chainlike manipulations of the tree (Fig. 1.6).

CSNB maps may strike the unfamiliar eye as strange, but these 'strange' figures are precisely the outlines necessary to be conformal, to show true shape. CSNB maps are analogous to, and as accurate and useful as floor plans are to architects; CSNB maps are no wilder than the 'wild' shapes of Maryland or Sulawesi, and no one would argue that a map of this state or that island should be warped to an oval or rectangle. Consider CSNB maps of Earth where the ridges and troughs of plate boundaries form the edge (Fig. 1.10). Save for marking these on a globe, there is no more exact representation than these, which account for all the real estate, both in extent and adjacency, and illustrate relationships between plates *in terms of* the boundary selection. Because the entire surface can be observed, and relationships demonstrated and measured simultaneously, the maps are arguably more revealing, more useful than their on-globe representations. Overlaying these with other maps, such as watershed boundaries, of the same object, or with analogous maps showing tectonic boundaries on other planets, is instructive, allowing similar or differing alignments, laps and gaps and their origins, to be precisely compared and appraised. The topic of interpretation will be discussed more fully in Chap. 3.

1.5 Relationship to 3D Modeling Techniques and Topology

Topology is the study of those properties of an object that, while they are in a sense geometrical, are the most permanent: the ones that will survive distortion and stretching. Topology is now a broad and fundamental branch of mathematics, with many aspects. CSNB is most closely related to combinatorial topology, originally known as *analysis situs*, which is the study of intrinsic qualitative aspects of spatial configurations that remain invariant under continuous one-to-one transformations. It is often referred to popularly as 'rubber-sheet geometry'. For example, deformations of a balloon that preclude puncturing or tearing are instances of topological transformations. A circle is topologically equivalent to an ellipse.

Topology has experienced an intensive evolutionary period for much of the twentieth century, incorporating progressively less geometry and more algebra and set theory, according to Brouwer (1913). Weyl (1913) emphasized the abstract nature of a surface as a two-dimensional manifold, a concept that should not be tied to a point space (in the usual geometrical sense), but given broader meaning, tying topography into set theory. The classical formulation of a set, E, of 'points' or elements, x, and their subsets, S, in neighborhoods, such as U(x), providing continuity in their relationship, was developed by Hausdorff (1914), father of point-set topology, from these axioms. Through additional axioms he developed the properties of various more restricted spaces, such as the Euclidean plane.

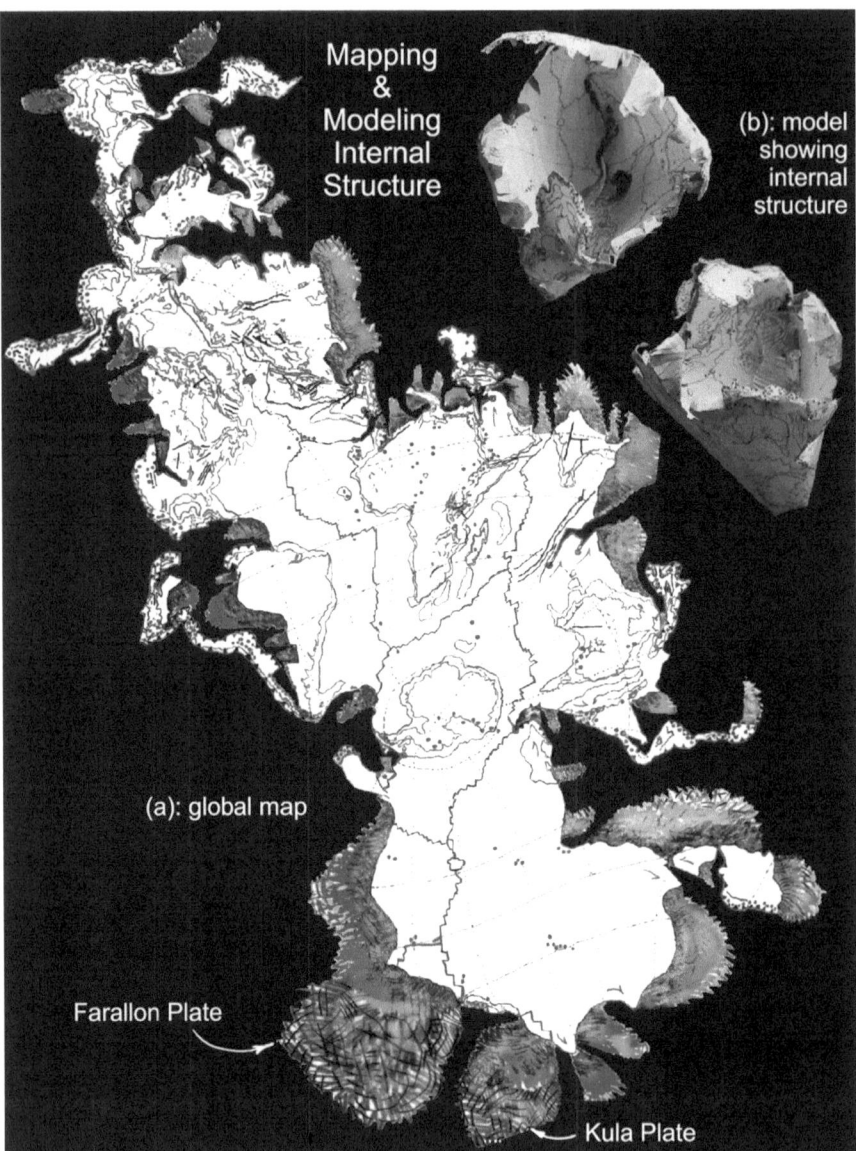

Fig. 1.9 Segmented plate boundary map of Earth folding into 3D model, as discussed in text (Clark 2004a, b) (Source data courtesy of NASA/GSFC, Global Tectonic Activity Map, Paul Lowman)

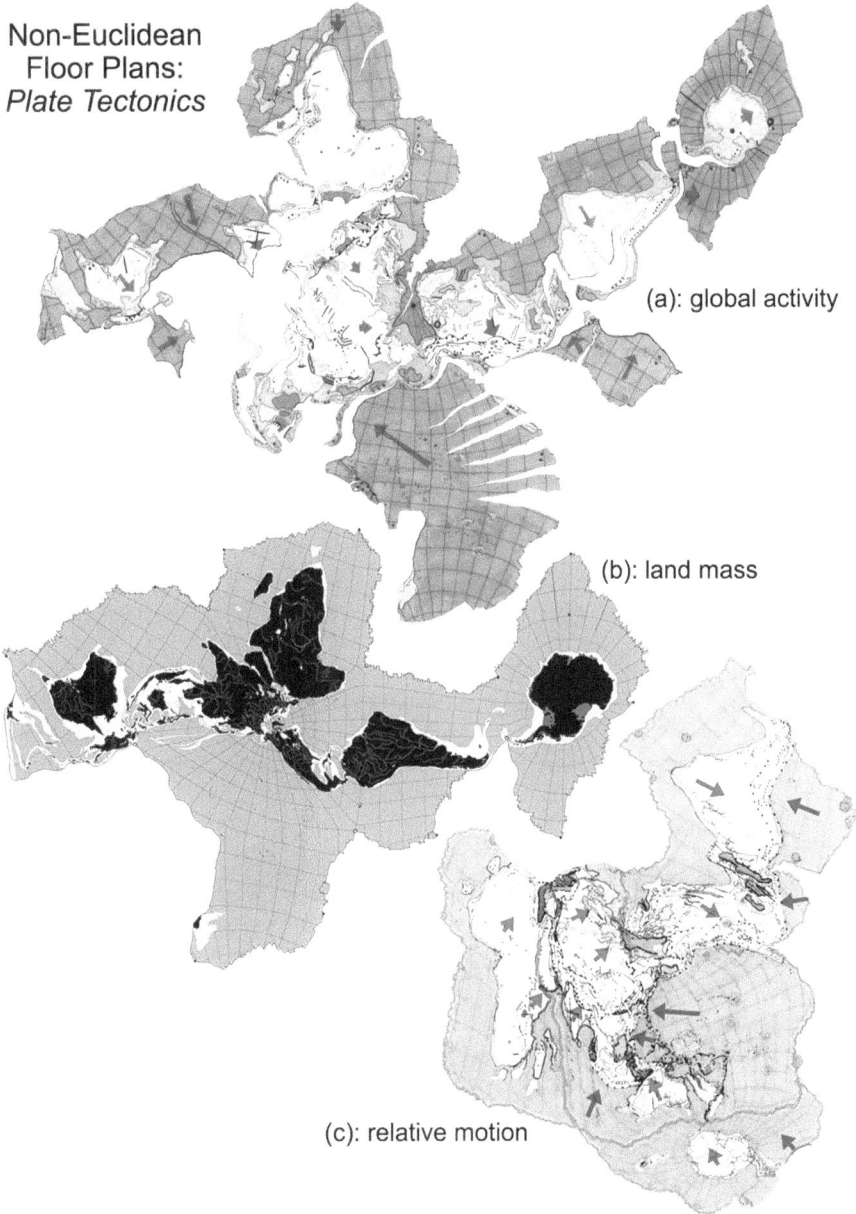

Non-Euclidean Floor Plans: *Plate Tectonics*

(a): global activity

(b): land mass

(c): relative motion

Fig. 1.10 Compactness progress in plate boundary maps (**a**) to (**c**) as discussed in text. Source data courtesy of NASA/GSFC, Global Tectonic Activity Map, Paul Lowman. (**a**) (Clark 2003); (**b**) With Jim Hagan, Gondwana Circle Design Competition (2010). (**c**) Suggested by Dave McAdoo

1. To each point x there corresponds at least one neighborhood U(x), and each neighborhood U(x) contains the point x.
2. If U(x) and V(x) are two neighborhoods of the same point x, there must exist a neighborhood W(x) that is a subset of both.
3. If the point y lies in U(x), there must exist a neighborhood U(y) that is a subset of U(x).
4. For two different points x and y there are two neighborhoods U(x) and U(y) with no points in common.

Although the word *point* is used in the concept, the new subject has as little to do with the points of ordinary geometry as with the numbers of common arithmetic. The concept of topology has emerged in the twentieth century as a subject that unifies almost the whole of mathematics, somewhat as philosophy seeks to coordinate all knowledge. Because of its primitiveness, topology lies at the basis of a very large part of mathematics, providing it with an unexpected cohesiveness (Boyer 1968). The portrait paintings of Chuck Close are an illustration of Hausdorff neighborhoods (Greenberg and Jordon 1998). Simple examples of such pixelization can be created with image manipulation software (Fig. 1.11). Each cell, artistically scrambled and apparently nonsensical, nevertheless contributes its proper share to the overall portrait when seen from a distance, and thus in context with its neighbors. This is the organization of pixels on a computer screen.

Topology and hyperspatial analysis became increasingly abstract between 1920 and 1960, providing the foundation for homological algebra, an abstract algebra generalizable and useful for pattern recognition in a broader range of domains, both natural and constructed (Boyer 1968), supporting hyper-abstraction as the language of post-modern science. Mathematicians (e.g., 'Bourbaki') have come to see mathematics as a "storehouse of abstract forms" that, for unknown reasons, empirical reality is "preadapted" to fit (Aczel 2007).

CSNB mapping is prototopological, harkening back to geometry-based topology concerned with the relationship between surfaces as planes defining overall shape. In pure topology, the middle has no significance: stretchability in the plane is endless and inconsequential, compared to the object's overall shape, as represented by the number of handles or holes, as in a doughnut or a coffee mug. The CSNB technique identifies consequential features, and maps these as planar, shape-defining boundaries in terms of processes driven by these features, or origins. This leads to better, or at least higher-dimensional, understanding. The black shape superimposed on Fig. 1.11's pixelated portrait is a CSNB map of a human head; the folded figure in the lower right demonstrates that the apparently wild outline, with no data in the field, nonetheless captures three-dimensional information. CSNB maps can represent forms that range in complexity, as roughness increases relative to size, from oblate, hill-and-dale spheroids to heavily faceted but continuous objects, e.g., asteroids, and beyond, to complex, multi-holed forms, as we show in Chap. 6.

3D modeling, heavily influenced by current topology conventions, involves the use of specialized software to create a representation of a three-dimensional object. 3D rendering is used to display such an object as a two-dimensional image. As in

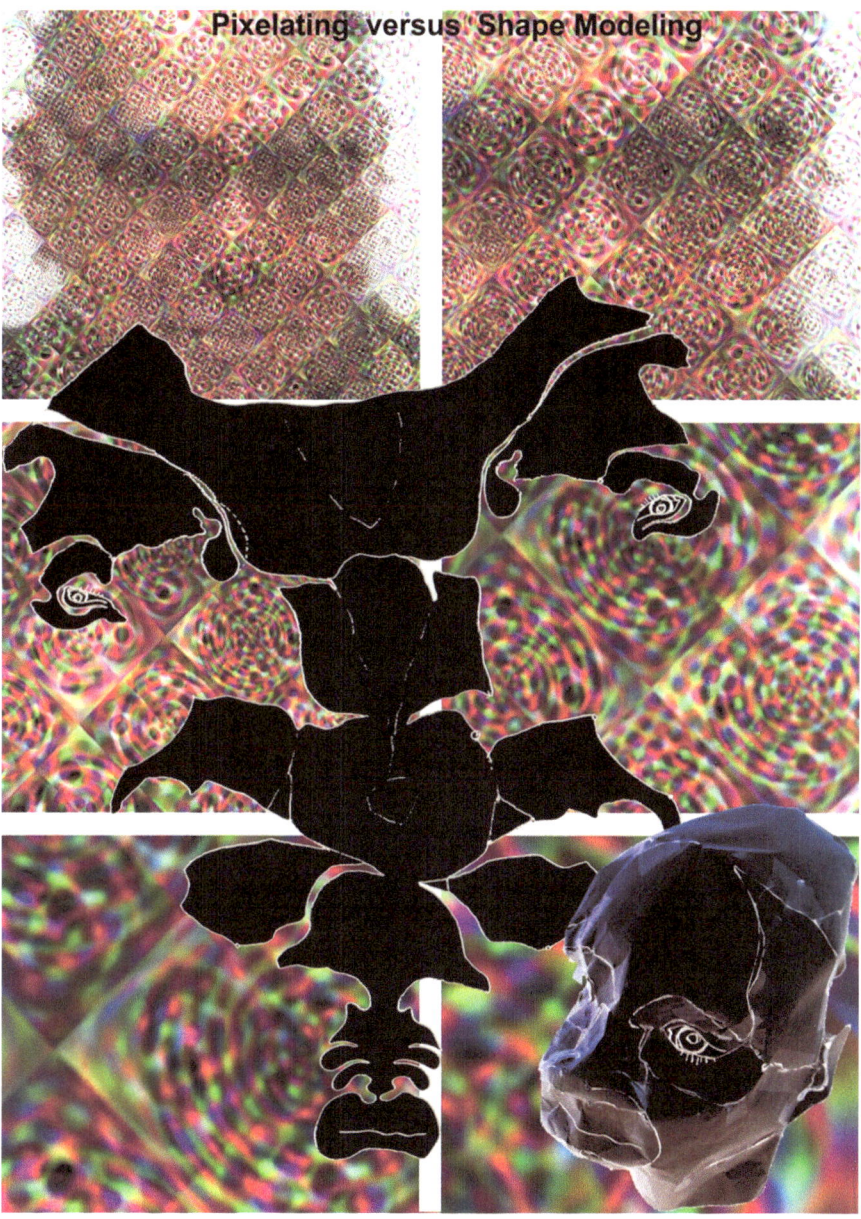

Fig. 1.11 Pixelation, at progressively higher resolution of Paul Lowman photo (see Dedication) based on Hausdorff neighborhoods (background, *top* to *bottom*) and a CSNB map of a head as discussed in text (Courtesy of C.S. Clark, "Pixelated Paul" © 2012 and "CSHB Human Head, Anonymous (Lee)" 2003, Photo by Sara Adkins Studio)

cartography, these processes involve 'shortcuts' in discrete digital approximations, or tessellation, which introduce distortions. Solid models define volume and internal structure using CAD techniques to provide realistic engineering or physical simulations and representations. CAD techniques involve constructive solid geometry and assume an object is composed of regularly shaped solid components. CAD surface models represent the appearance of a surface in three dimensions, but non-uniquely represented by a network of space-filling polygonal meshes, typically triangles (Fig. 1.12). As in conventional cartography, use of the arbitrary grid creates distortions, particularly wherever natural irregularities occur, and makes the establishment of relationships between globally distributed features and terrains more difficult. The CAD approach provides no basis for a link to a 2D projection or map of the same object where everything can be seen at once; an entirely different set of algorithms and assumptions must be used to access a map.

We note also that when conventional 3D models are used, discovery of global patterns in surface feature details from rotating dynamic views is problematic. Working memory, which relies on visual cues, has limits (Miller 1956). Global awareness must thus rely on numerical and statistical analysis, without facilitating mental synthesis. The difficulty in unscrambling globally distributed terranes and features, such as the Meridiani sedimentary record on Mars, illustrates this point.

1.6   Relationship to Perspective-Based and Anamorphic Drawing

CSNB method incorporates perspective techniques developed during the Renaissance, including the use of the vanishing point as described above, as well as nonlinear perspective techniques associated with anamorphic drawing (Fig. 1.13). Renaissance artists, including da Vinci, became very sophisticated in the use of both linear and nonlinear perspective (Veltman 1986). They observed that surfaces undergo apparent distortion in shape that enhances the perception of depth and changes the recognition and awareness of such objects as the position of the vanishing point shifts from the horizon, or recedes from (or approaches) the station point. Surfaces parallel to the picture plane diminish without distortion and with greater depth, whereas those not in the plane diminish with distortion and may no longer be recognizable. Renaissance artists created works to be viewed from multiple perspectives, as indicated by the creation of paintings where the eyes or the subject apparently follow the viewer (Wollaston 1824). They codified rules of perspective, as they measured and derived geometrical relationships as part of planning and executing a drawing, and clearly saw the relationship between the use of perspective in drawing, surveying, and optics (Vignola 1583; de Vaulezard 1630). Artists began to deliberately introduce distortion and depth, by, for example, choosing to position the vanishing point at other positions to emphasize or deemphasize certain features (van Thiel 1969). In fact, a scene could be deliberately drawn for viewing

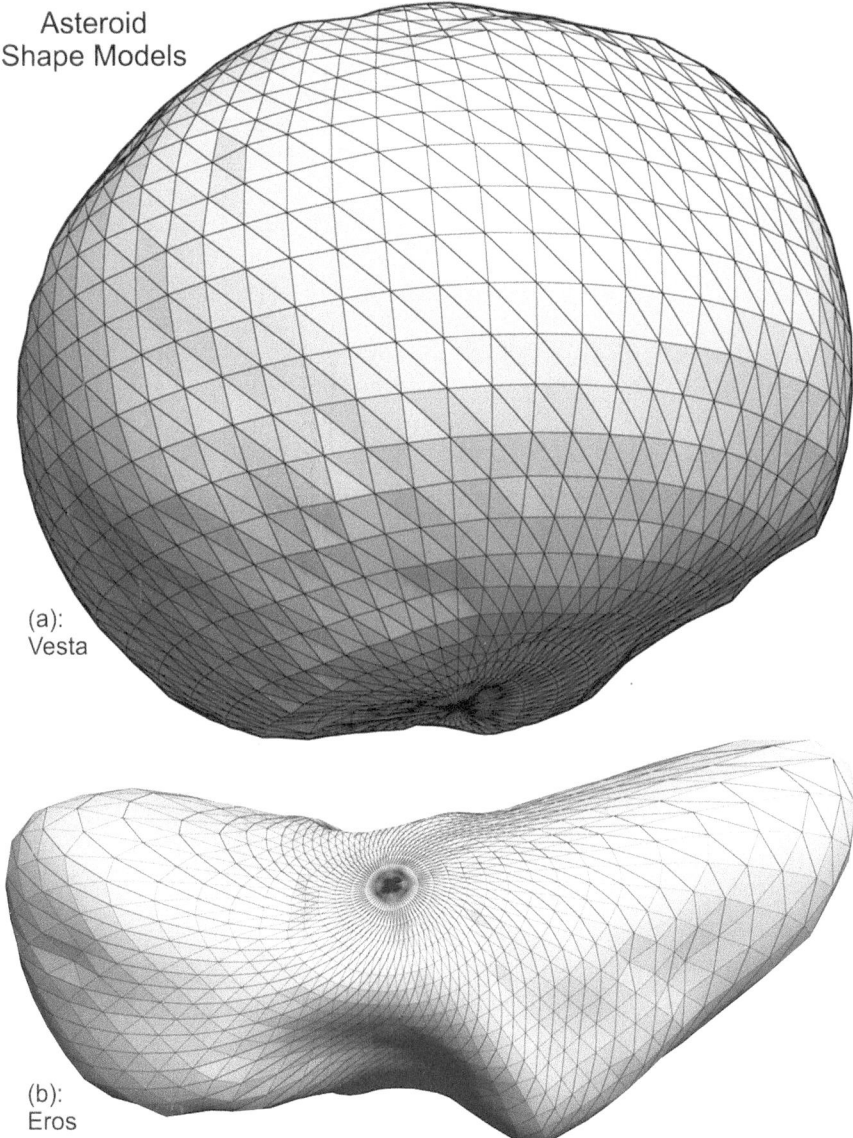

Fig. 1.12 Three-dimensional surface-covering meshes for relatively (**a**) regular asteroid Vesta and (**b**) irregular asteroid Eros. Images courtesy of Tayfun Öner

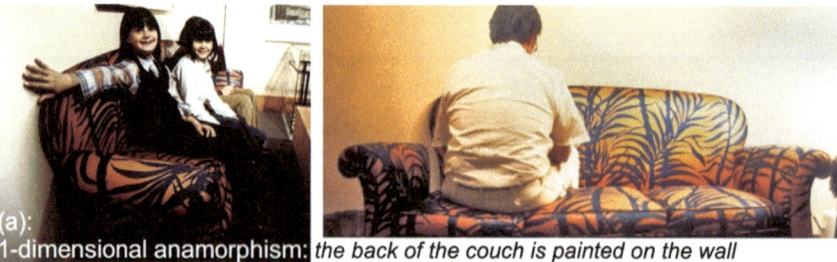

(a): 1-dimensional anamorphism: *the back of the couch is painted on the wall*

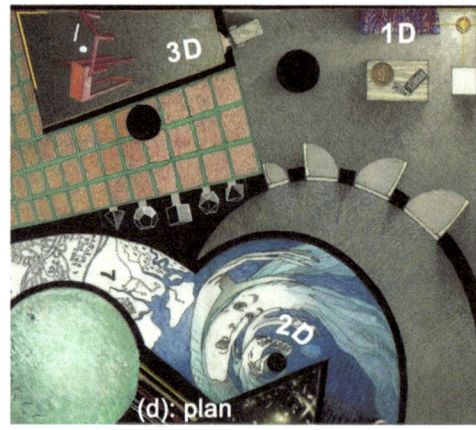

(d): plan

(b): 2-dimensional anamorphism: *floor and ceiling paintings are reflected in mirrored cones*

CSNB is a 4-Dimensional Anamorphism

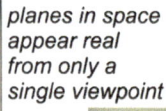

(c): 3-dimensional anamorphism: *planes in space appear real from only a single viewpoint*

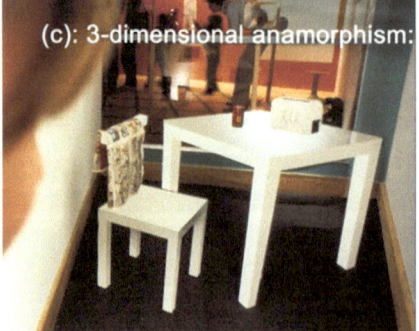

Fig. 1.13 (**a**) Nonlinear perspective, (**b**) conic rays intersecting a plane, (**c**) plane projections in space, relative to a unique station point, and (**d**) Plan reference. Sources artwork "Spaces and Illusions" at High Museum of Art, C.S. Clark (1977) (**a**) with L T Jones, (**b**) with Susan Clark Williams, (**c**) suggested by Merrill Elam, and (**d**) coordinated by Mack Scogin

from an extreme perspective, far from the 'normal' perspective of the viewer directly in front, with the vanishing point at eye level. This is referred to as anamorphic drawing (Marolois 1614).

da Vinci made the link between Euclidian geometry and anamorphic transformation, and even suggested the use of a pinhole camera or peephole to control the viewing angle for large, complex scenes. Over the next three centuries, mechanical and grid systems that used conical or cylindrical mirrors for drawing automatically from nonlinear perspectives evolved (de Vaulezard 1630; Leupold 1724; Baltrusaitis 1984; Barbaro 1569) Although these devices were designed to deal with the shifts of perspective to the left or right, and shifts above or below eye level were still treated separately, the use of spherical systems, which would treat shifts in any direction, were considered. Artists have never embraced the use of linear perspective-based techniques exclusively, instead opting for the use of linear perspective for landscape and building elements, and treating human subjects with nonlinear, optical adjustments (Veltman 1986). By the nineteenth century, vision theory had moved in a different direction. The constructivists (von Helmholtz 1924) broke its link to observations from which perspective methods were derived, separating the theoretical model of vision from observations and its practical use. The current computation approach relies on mathematical modeling to reconstruct physical scenes from retinal images (Marr 1982).

CSNB method uses nonlinear perspective and a form of spherical, as opposed to merely cylindrical, linear projection as a basis for locating an object's benchmark surface points, e.g., the poles, on the map. Boundary segments (at constant scale) are analogous to plane traces, i.e., lines with a common intersection, or *hinge*, not unlike the vanishing point of a roof valley in a two-point perspective of a cross-gable house. Thus, the CSNB approach could encourages the unification of scientific understanding of physical processes with observation in the service of making maps to illustrate physical processes.

Chapter 2
Constant-Scale Natural Boundary Mapping Technique

2.1 Identifying Critical Boundaries, Unzipping and Zipping

The 'divide tree' or 'tree of interruption' is a network of boundaries that control a CSNB map. It grows from trunk lines of topographic extremes (ridges or valleys) identifiable on a shaded relief or topography map as strings of nearest neighbor summits (maxima) or pits (minima). For irregular objects, these can be thought of as maximum angular inflexions of planar orientation relative to the center of mass. Generally, the tree grows in either top down (from maxima) or bottom up (from minima), starting with the most recognizable and extreme cluster of maxima or minima. However, if the intent is to consider a certain class of features, such as plate boundaries, only maxima or minima associated with these boundaries should be used. The next step involves flattening: the transformation of the generated tree (and its surrounding area) from the spherical to the planar surface. See the example in Fig. 2.1.

Boundaries never change in length, only in orientation; boundaries define regions, and hence centerlines of regions. Centerlines are the pruned trunks of medial axes, which we discuss in Sects. 2.6 and 6.5. The medial axis is the set of center points of all circles equidistant from surrounding boundaries. It can be useful in locating centroids, direction vectors, and measures of distance in preparation for the next step. Centerlines change only in length, never in orientation. If these restrictions were relaxed, and the edge could be any imagined and stretchable line (such as a line of longitude), then conventional, formula-based projections would result.

How far the boundary tree grows depends on data resolution as well as preference for map segmentation or compactness to best illustrate processes. The splitting of boundaries, which creates segmentation, is known as *unzipping*, the reverse process, which creates compactness and a continuous surface as *zipping*. Highly segmented maps with largely unzipped boundaries illustrate morphological differences and patterns within an object, whereas compact maps, largely zipped, illustrate a global overall shape.

P.E. Clark and C. Clark, *Constant-Scale Natural Boundary Mapping*
to Reveal Global and Cosmic Processes, SpringerBriefs in Astronomy,
DOI 10.1007/978-1-4614-7762-4_2, © The Author(s) 2013

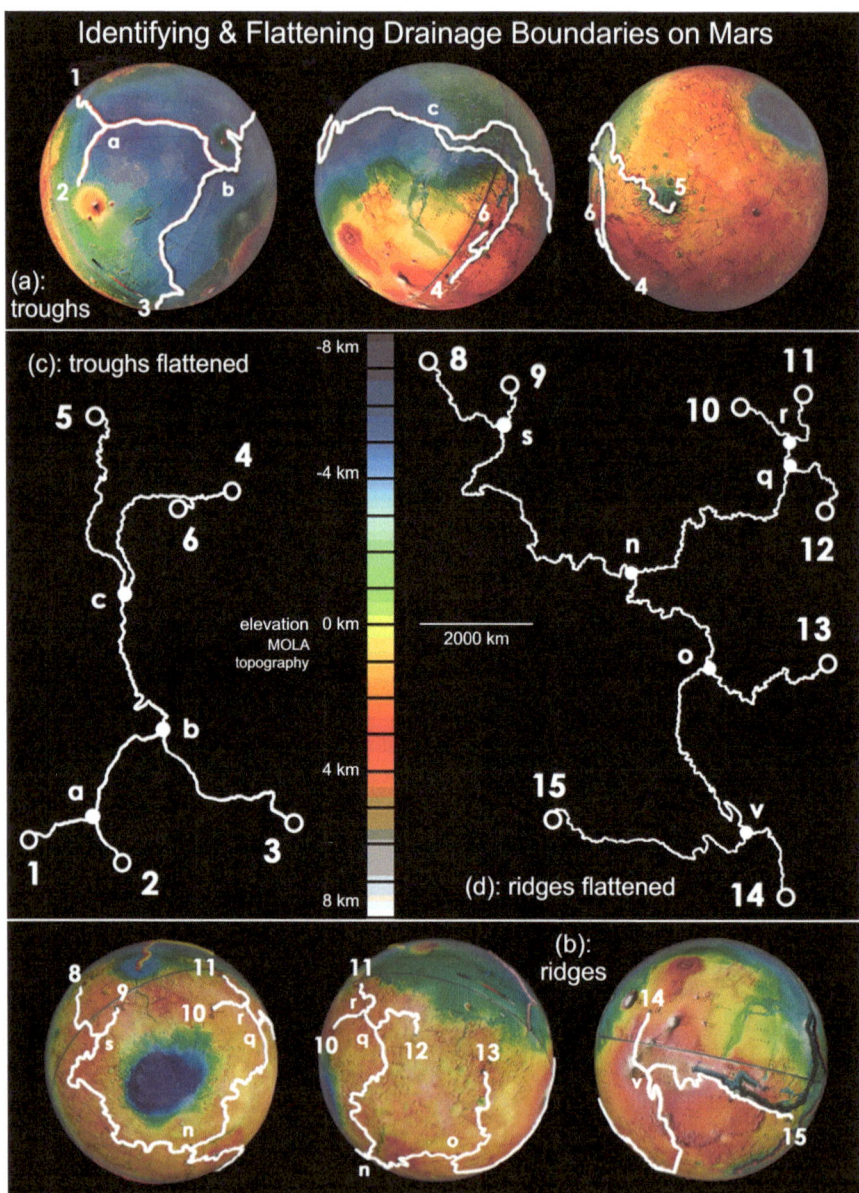

Fig. 2.1 CSNB mapmaking using (**a**) Martian trough and (**b**) ridge constant-scale boundary selections, then (**c** and **d**) flattening and hinging, as described in text (Clark 2004b) with help of René De Hon (Source data: Mars topography (MOLA) data (*red high* to *purple* low elevation). Courtesy of NASA)

2.2 Making Closed Shapes and Adjusting Proportions

We arrive at CSNB's second step—restoring a bounded surface.

A sphere's unbounded yet finite surface becomes a plane's unbounded, infinite surface as boundaries transform via radial unbending. Interruption of the surface happens by 'unhooking' one branch tip and unzipping the boundary along its length; additional tips also hinge. Rejoining the loose ends soon binds the once-infinite plane, reversing Fig. 2.1's relationship, as seen in Fig. 2.2.

It is not immediately clear that overlaps may always be eliminated, but pruning branches and altering hinge-arcs has always worked. In early CSNB maps (1992–1998), hinge-arcs were managed according to either of two simple criterions:

1. Equal or nearly so, thus equalizing distortion near hinges. Early maps, e.g., Figs. 1.3a and 1.8, were segmented, and conformality intuitively obvious. This strategy kept area and shape distortion to a minimum.
2. To maximize enclosed area. This tactic evolved as compact maps were attempted, e.g., Fig. 1.7a. (Also seen in Fig. 2.5 discussed in Sect. 2.4; note shape distortion near Hinge 6.)

Tracking only hinge-arcs inevitably created skewing in the reconstituted shape. Hinges must be set at a collective optimum, which exists, and may be found by correlating lengths across the flat shape with corresponding lengths on the object's spherical or irregular surface. If we, as Dürer, were mapping polyhedrons, we would make map lengths *equal* to corresponding object lengths (shown in Sect. 2.7). For natural bodies, we make map lengths *proportional* to corresponding object lengths.

Figure 2.3 shows this adjustment for the Mars trough-bound map. White lines are cross-map lengths, which are adjusted by equalizing their several ratios as a group, as if they were strategic lengths of springs. Hinges rotate in response; balanced spring strain ensures correct mid-map proportions. Where a suitable hinge is unavailable, as between Points 1 and 5, secondary bends, called *elbows*, are introduced between adjacent hinges. (Fig. 1.7 shows construction of a conformal elbow.) Elbows compromise a perimeter's angular metes and bounds, and thus degrade proportions, but the effect is slight, and local. Similarly, proportions will be inevitably distorted near hinges, but this effect is also localized.

A key distinction from, say, a circular or elliptical global map is the use of as few elbows as possible. This minimizes conformal distortions. The trade-off is in CSNB's extravagant shapes, which, it turns out, are precisely the shapes required to accurately capture antipodal proportions. This becomes clear as we cultivate our map's interiors.

2.3 Adjusting Internal Scale

In conventional cartography, scale is a function of latitude and longitude. In CSNB mapping, the interior points form a curving surface surrounded by a flat, constant perimeter. The interior surface points are projected to the plane by the two ancient

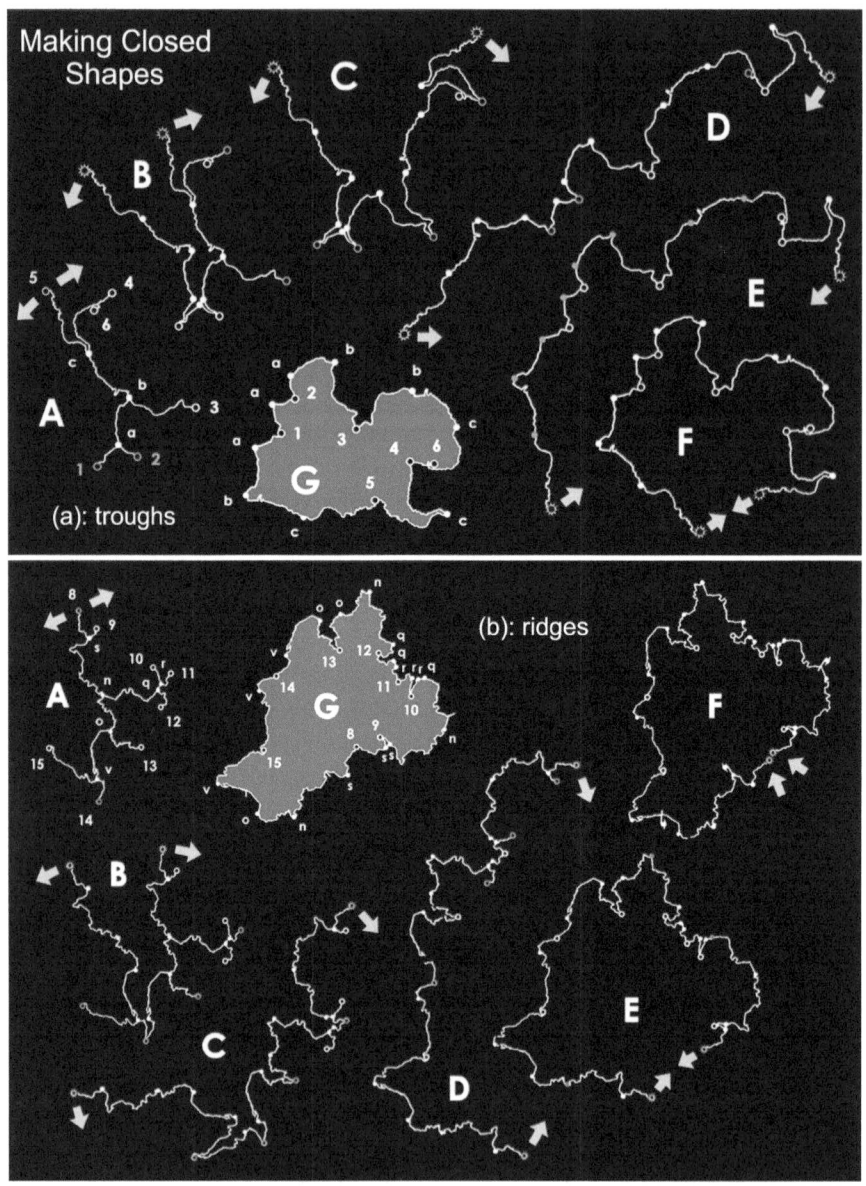

Fig. 2.2 Making a closed shape (using map in Fig. 2.1) via unzipping and hinging, as described in text Clark (2004b) with help of René De Hon

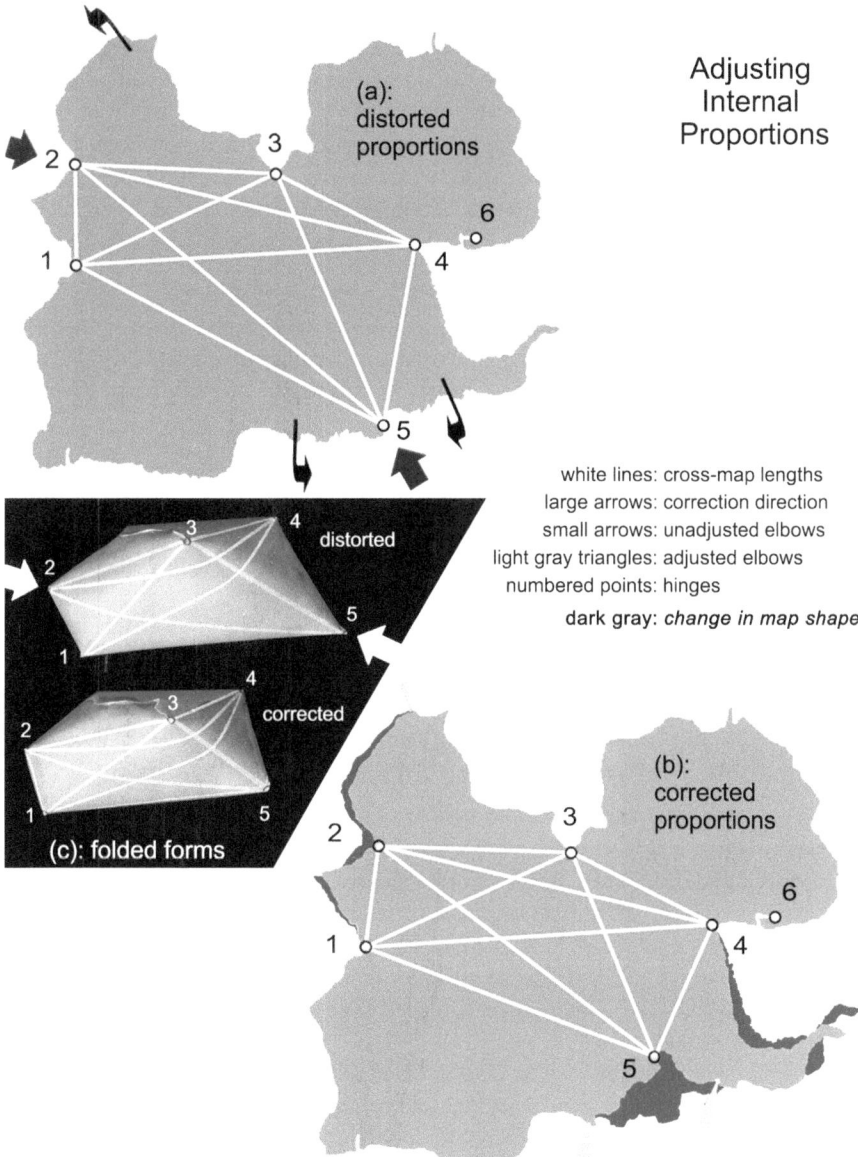

white lines: cross-map lengths
large arrows: correction direction
small arrows: unadjusted elbows
light gray triangles: adjusted elbows
numbered points: hinges

dark gray: *change in map shape*

Fig. 2.3 Adjusting internal proportions (using map in Fig. 2.1) as described in text. Clark (2004b) with help of René De Hon

methods of geometry attributed first to the mapmaker Ptolemy (Sidoli and Berggren 2007) orthographic and stereographic projection (Fig. 2.4a). In an orthographic projection, the scale varies as smoothly as on a conventional map. Continents and oceans are both reduced evenly in size from their global dimensions. In a stereographic projection, scale may be specified as a function of distance from a given point, feature, or region, as in Fig. 2.4b, which concentrates areal distortion along mid-ocean ridges in order to show land at constant scale with the edge, i.e., continental size is preserved and ocean crust contracts in visual mimicry of a tectonic look backwards in time.

2.4   Drawing the Grid and Creating a Map

To provide orientation on maps and on surfaces of globes, space is usually subdivided into a grid of latitude and longitude, called graticles. Conventional map interiors are a precise and unambiguous coordinate point system within an arbitrary perimeter and graticles derive from a projection formula. CSNB interiors are controlled inexactly, as groups of points within a precise, unambiguous perimeter, and graticles are derived by adjacency, e.g., "neighborhood" axioms of Hausdorff (1914), with two added stipulations, the first obligatory, the second (usually) desirable:

1. Size of the smallest (most central) neighborhood is set arbitrarily, and
2. Sizes of other neighborhoods smoothly vary with distance from the edge.

Once this step is taken, the CSNB map has been created (Fig. 2.5).

We could use, and have used, a trial-and-error method for adjusting internal proportions. Graticles could be sketched in, compared to their on-globe counterparts, distortion appraised, and hinges adjusted to reduce skew. However, the approach outlined in Sect. 2.2 appears to allow more rapid minimization of distortion.

The development of the boundary network can begin on a conventional projection map. However, boundary lengths will need to be adjusted for their actual length to maintain constant scale.

Drawing natural boundaries on a flat, conventional map provides an interesting contrast to a CSNB map, particularly if the conventional map is reconfigured so that the natural boundaries form its edges (Fig. 2.6).

A succession of bordering lines may locate points in a map's field by organizing the map into a series of concentric neighborhoods. These "Hausdorff waterlines", which will be discussed in Sect. 2.6, move inward from the map edge a uniform distance y_1 y_2 y_3 ..., and subdivide the field into neighborhoods of points relative to the border. Points in the inmost neighborhood(s) are, obviously, those points most distant from the edge, both on the map and on the globe.

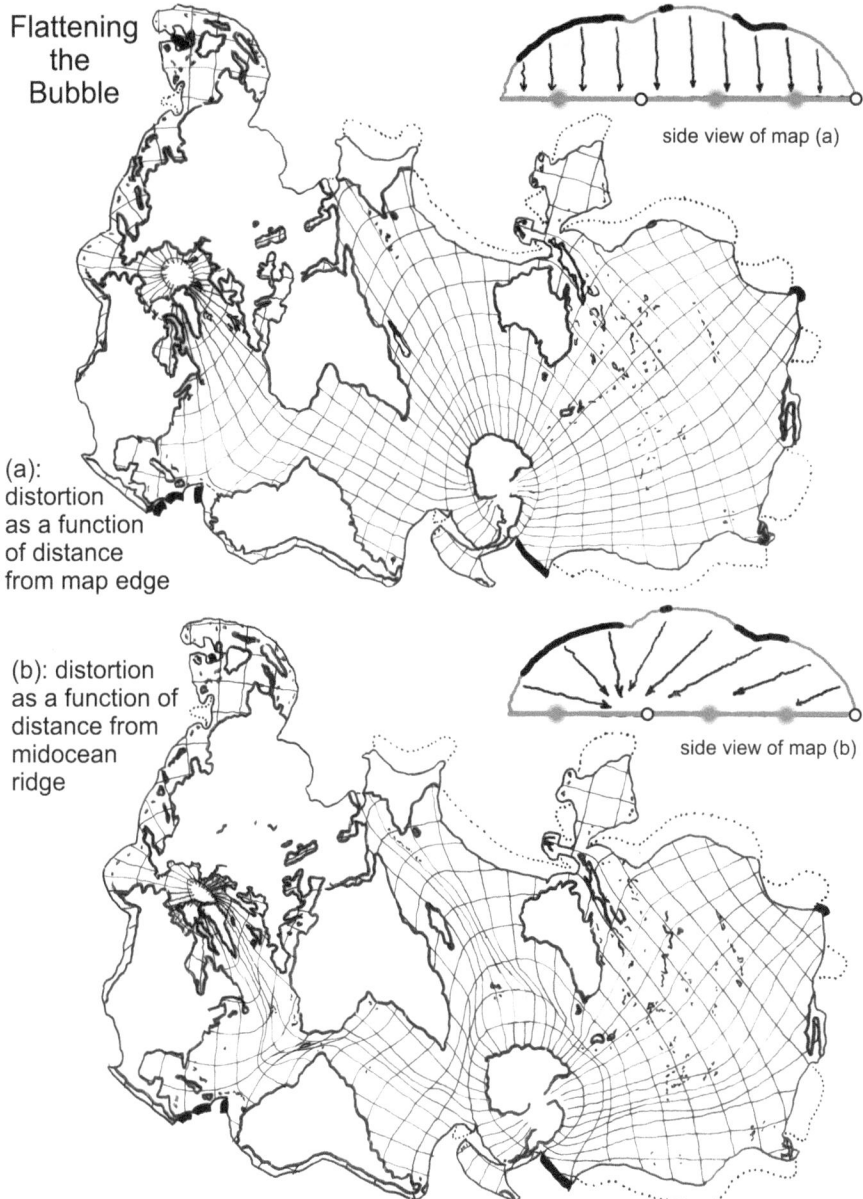

Flattening
the
Bubble

side view of map (a)

(a):
distortion
as a function
of distance
from map edge

(b): distortion
as a function of
distance from
midocean
ridge

side view of map (b)

Fig. 2.4 Flattening the bubble method drawn by Chuck Clark in 1997, based on a concept from Ptolemy, for dealing with distortion, as described in the text

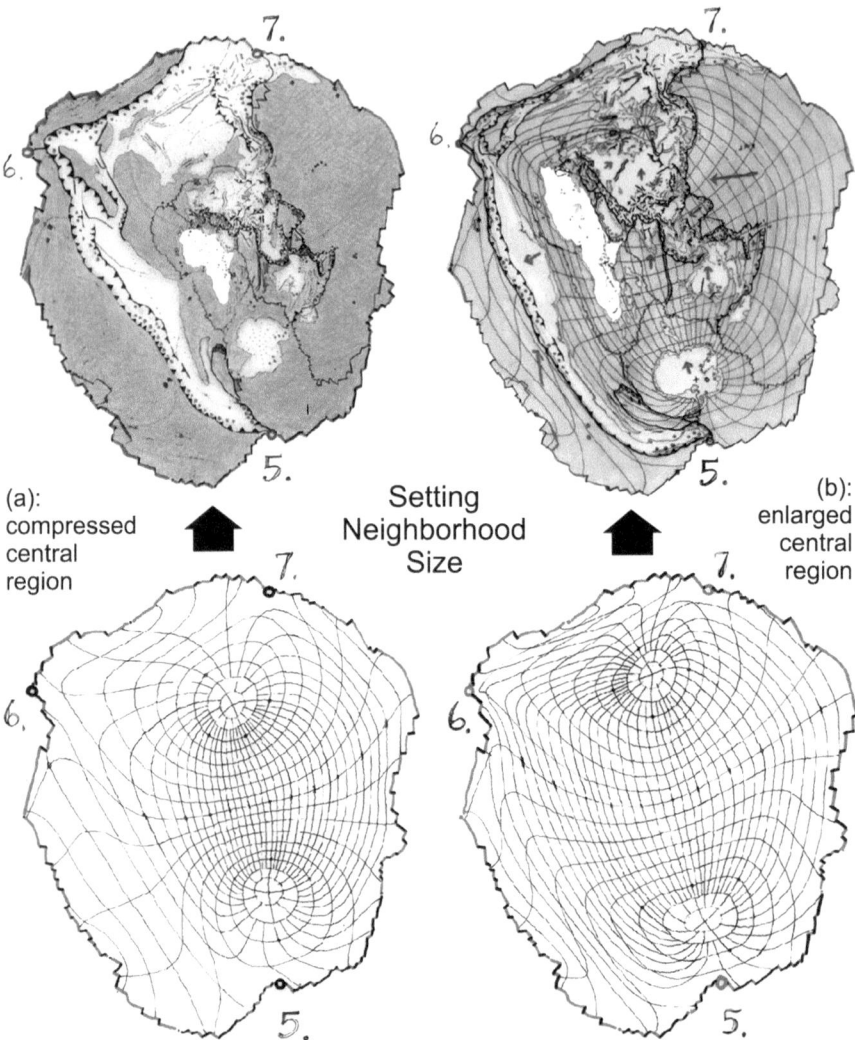

Fig. 2.5 The graticle-grid (*bottom*) used to create CSNB maps (*top*) (Clark 2003). All maps have constant-scale perimeter, but *left* and *right columns* illustrate different algorithms for varying distortion (compressed versus enlarged central region as seen on bottom) as a function of distance from perimeter. Note differences in neighborhood size and location (e.g., Africa) (Source data: Courtesy of NASA/GSFC, Global Tectonic Activity Map, Paul Lowman)

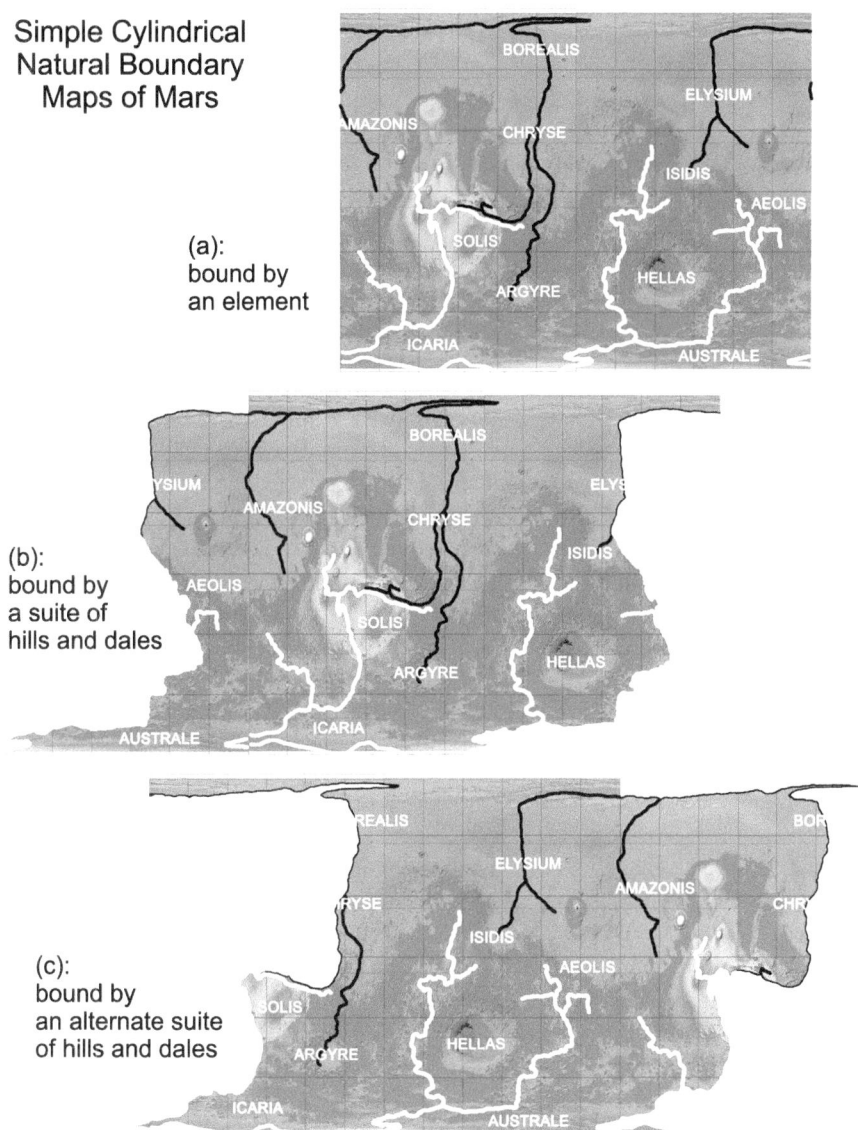

Fig. 2.6 Recomposing Mars conventional maps to represent natural boundaries. Spilhaus (1991) used this method for terrestrial maps. Clark (2004b) with help of René De Hon (Source data: MOLA courtesy of NASA)

2.5   Folding

We can test the accuracy of CSNB maps by folding into a 3D model, as illustrated in Figs. 2.3 and 2.7. Folding is a unique property of this approach and signifies the ease of 3D-2D-3D transition. CSNB maps should fold neatly, without gaps. If not, a mistake exists in the map's border. Forms should 'close' in a similar sense as an accurate survey does when metes and bounds are plotted.

We see a larger inference than mere drafting accuracy in this property. The lower form in Fig. 2.3c is a smaller volume because of its smaller surface. While it shouldn't surprise topologists that CSNB maps always fold, the volumetric change within a constant perimeter, Panofsky's prototopological property, demonstrates that constant edge scale is the sole necessity for foldability. This is dramatically illustrated in Fig. 2.7: a CSNB map made from a fine-branched tree of severely limited global extent is accompanied by its bizarre but nevertheless neatly closed folded form. Unlike conventional maps, CSNB maps have conformal antipodes, and unlike geodesic maps, which also fold, CSNB antipodes are naturally bounded. CSNB folded forms are not planar-faced, straight-line-edged solids, i.e., polyhedra; instead, they have one, variable and irregular face with multiple vertices. These solids may be novel geometrical entities. By inspection, they appear to be self-organizing amalgams of cones, cylinders, and twisted sheets. We might call such solids *unihedrons* or *unitopes*.

When CSNB maps are partially folded, exterior or interior appendages, such as submerged tectonic plates, can be modeled, as we saw in Fig. 1.9. And when CSNB maps are completely folded, they have weak and strong axes, which may mimic an object's actual conditions, such as Earth's trenches and spreading ridges. We envision adjustable mechanical globes with interlaced and manipulable boundaries, able to demonstrate the push-pull of plate tectonics. (An example is in Sect. 4.2.)

For regular objects, the only true 3D representations are triaxial ellipsoids, so such models, while instructive in showing relationships, will not be true representations. However for irregular objects such as asteroids, CSNB models made from highly segmented maps will be the truest representations, as discussed in Chap. 6. And printed flat sheets can be folded to make 3D photoreal models, a capability of obvious economic virtue, a dollop of public fascination (Jacobs 2009) and high educational value (Lakdawalla 2008).

2.6   Waterlining

The nature and impact on selected intra-plate stress fields generated by tensional or compressional physical processes at boundary lines or points can be captured semi-quantitatively by using the CSNB waterlining technique (Fig. 2.8). To waterline is to draw a series of concentric lines paralleling an initiating line (Blum 1967); such lines represent stress contours (Zoback 1992; Christensen 1999; Siddiqi and Pizer 2005).

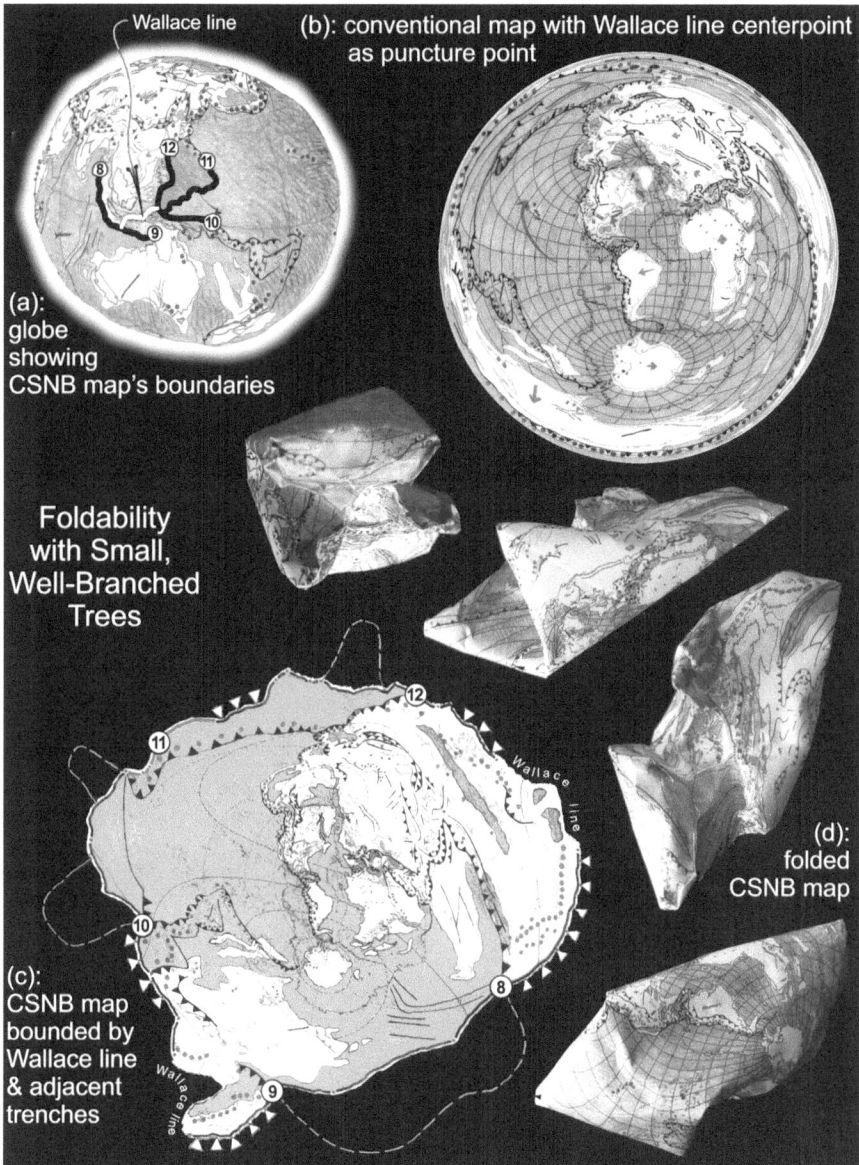

Fig. 2.7 Earth from the viewpoint of Wallace's line (Whitmore 1981), a boundary based on species distribution, and on nearby tectonic activity, as discussed in Sect. 4.2 (Clark 2004a). Even a localized, well-branched tree generates a map and folded form, though the result is bizarre (Source data: courtesy of NASA/GSFC, Global Tectonic Activity Map, Paul Lowman)

Waterlining

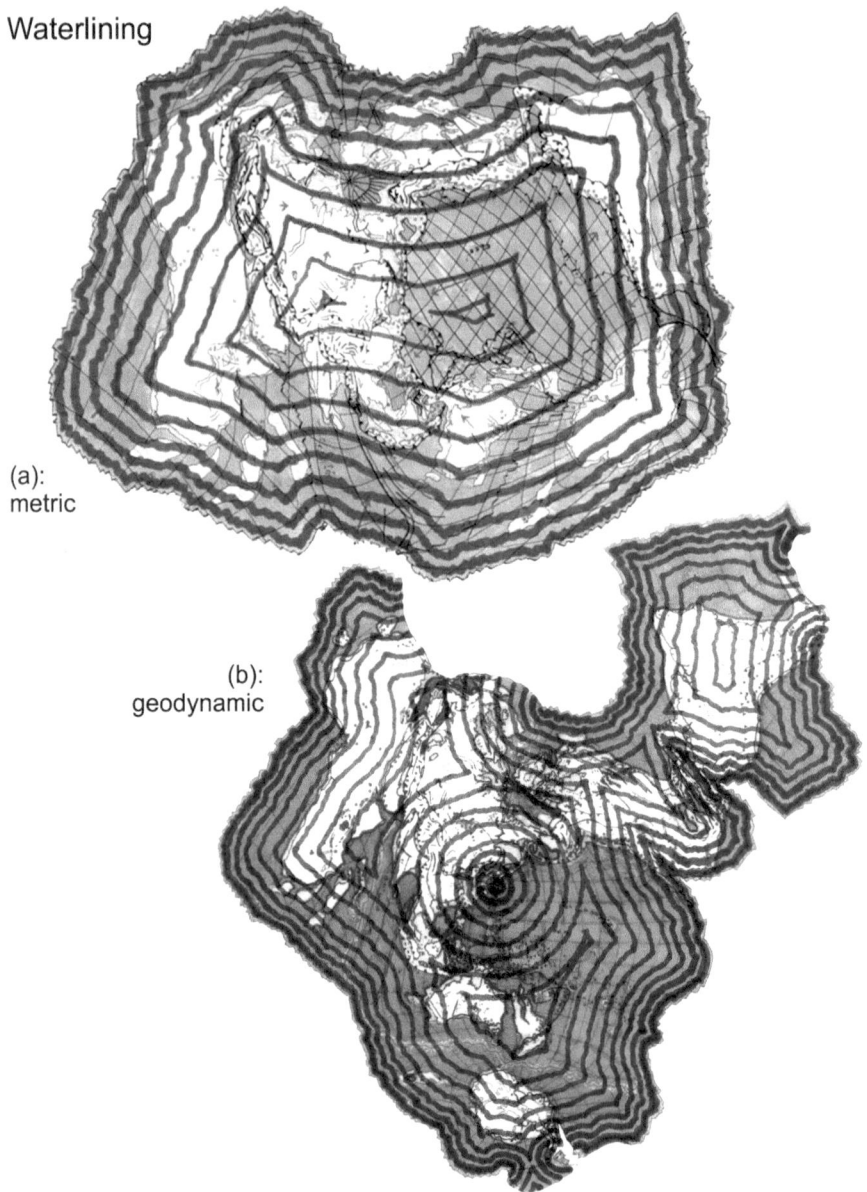

(a):
metric

(b):
geodynamic

Fig. 2.8 Illustration of (**a**) metrical and (**b**) geodynamical waterlining as discussed in text (Source data: courtesy of NASA/GSFC, Global Tectonic Activity Map, Paul Lowman. Drawn by Chuck Clark)

The boundary may be thought of as a stimulus, an initiating wave; hinges may be thought of as first order, mid-plate, uniaxial stress vectors best fit to the data (Lowman 1999). First order stresses express as inward wavefronts, and second order stresses as functions carried on the first order waves. Hinge rotation and location could be adjusted to model observed stresses.

Similarly, a series of concentric lines representing stress movement may be drawn around a single, point-like event occurring anywhere on the map. The resulting interference patterns may reveal previously undiscovered associations, and show correlations with terrane types or features, as suggested in Fig. 2.8. One set of waterlines works inward from the map's edge (a segment of the mid-ocean ridge system), and expresses the effects of sea floor spreading (metrical) (Fig. 2.8a). A second set of waterlines radiates outward to express the effects of the 2011 Tōhoku megathrust earthquake (geodynamic) (Fig. 2.8b).

The medial axis can be useful in locating centroids, vectors, and distances. Waterlines can be used to locate the medial axis, as in Fig. 1.8, where ocean-basin labels follow a medial axis established by waterlines radiating from shore. The medial axis may include or be merged with fractures, faults and trenches, and wavefronts or critical boundaries redrawn accordingly. Inversely, the medial axis, a topological skeleton akin to a CSNB tree, could be portrayed in the greatest detail, and its stresses, including Coulomb stresses associated with earthquakes, emphasized by becoming critical boundaries, a CSNB map edge.

2.7 Demonstration

A pipe cleaner demonstration (Fig. 2.9) is helpful in visualizing the CSNB method. In moving pipe cleaners from globe to table, the only geometry in play is *development* and not, as in conventional maps, *projection*. A sphere's finite, unbounded surface becomes a plane's infinite, bounded surface, as the pipe cleaners leave the grapefruit and flatten on the table. At A, the object's surface has not yet been interrupted. This happens at B by unhooking one pipe cleaner and unzipping the boundary along its length (CDE). Reconnecting the pipe cleaner soon binds the plane's infinite field. At this stage, while nothing is required *in* the field of the map, a shape model exists, as seen in the folded forms (as we saw in the foreground of Fig. 1.11).

Figure 2.10 illustrates CSNB mapping of a regular polyhedron, in this case a brick. A tree marks its edges. The tree is fully inter-digitated: if a branch were lopped, the result would be deficient; if it grew another branch, a circuit (prohibited in this simplified exposition) would occur. As we saw in Fig. 2.1, the sequence ABCD shows the flattened tree unzipping, branches swinging to enclose a shape. Sequence D´EFG shows the shape's internal proportions adjusting via diagonal distancing (dashed lines). Shape D is, of course, the 'developed geometry' begun by Dürer. Shapes DEF fold to forms DEF, which are prototopological equivalents to the brick; form G is the brick's geometrical replica.

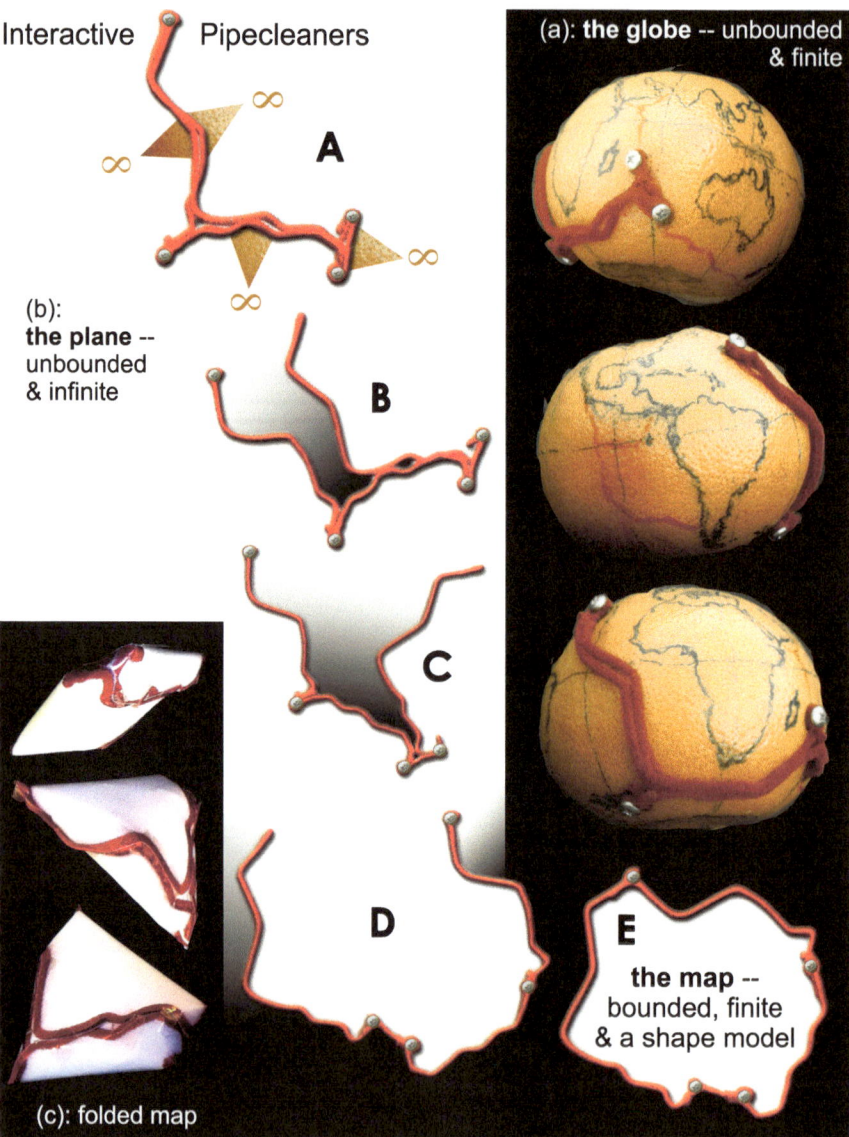

Fig. 2.9 (**b**) Demonstration of developing, unzipping, hinging, and reconnecting boundaries (Clark 2003), (**a**) their relationship to original object, and (**c**) 3D shape model. Note that (**c**) is prototopologically equivalent to (**a**), given the selected boundaries. From demonstration at Advances in Extraterrestrial Mapping workshop (2003)

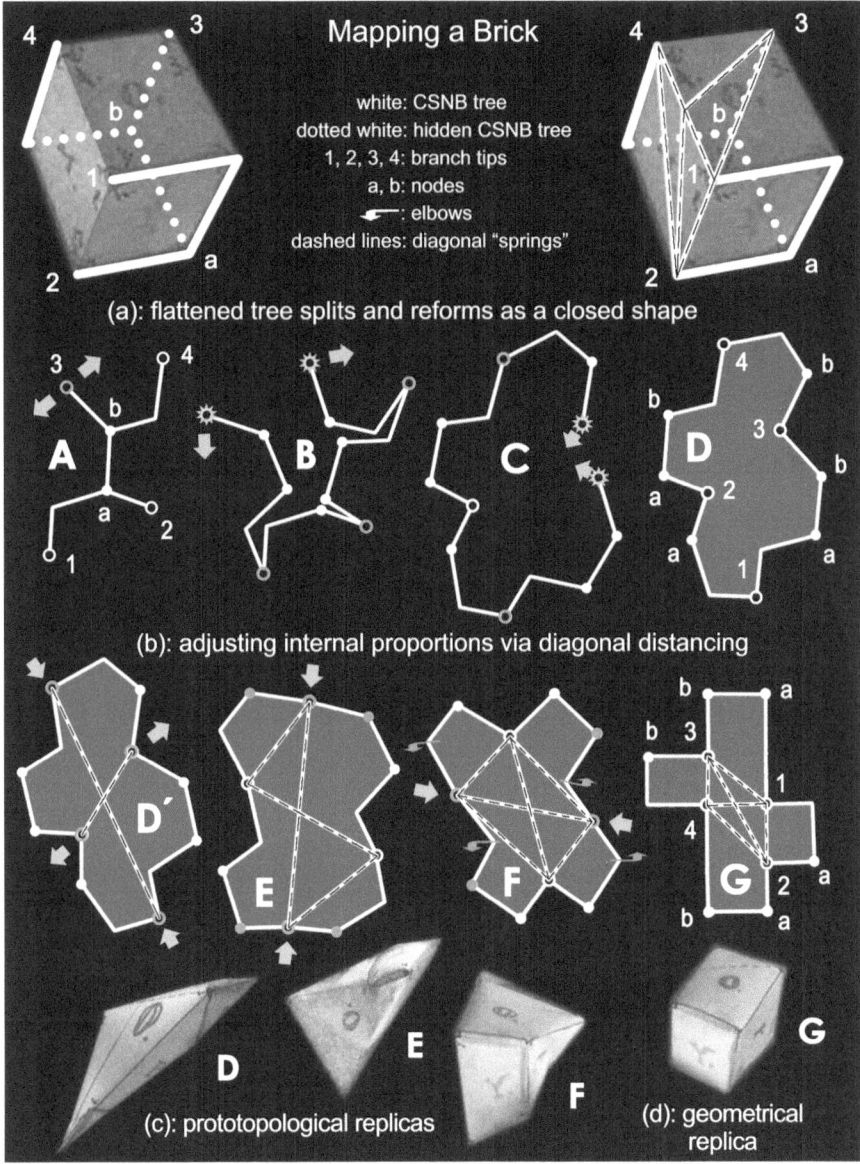

Fig. 2.10 Step-by-step CSNB mapping of a brick, as discussed in text (Clark 2007) (Suggested by P.E. Clark and Gunther Kletetschka)

2.8   Summary of Implications for Global Mapping

CSNB maps have these distinctive features:

1. The map edge is a deformable wavefront, with resolution dependent on boundary complexity.
2. Models, folded along edges, are condensations of the object, and, for irregular solids, such as asteroids, can be true representations.
3. Resolution is highest at the most recognizable, pivotal, and defined feature, forming the boundary, where local proportions are preserved.
4. The CSNB approach, unlike conventional map projections, preserves antipodal geometry (maps are conformal for antipodal areas).
5. Local interior proportions are also preserved. Equalizing ratios of hinge connector cross-map-lengths to lengths between corresponding object points minimizes shape distortion.
6. As resolution increases and/or more natural boundaries are discovered, additional edges subdivide and refine the original shape model.

Chapter 3
Interpretation of CSNB Maps

3.1 Nature of Processes and Resulting Boundaries

For a given body and timeframe, internal (volcano-tectonic, chemical, or thermonuclear), external (erosional/depositional or rotational), or combined processes may dominate in shaping global maxima or minima surface features, which act as constant-scale boundaries and terrane edges. CSNB mapping has now been used to produce global maps of bodies lying on a continuum between externally and internally controlled surface morphology (Clark 2002, 2003, 2004a, 2004b, 2005, 2007, 2011; Clark and Clark 2005, 2006a, b, 2007, 2009, 2010; Clark et al. 2006, 2007).

Surface morphology is the historical expression of combined internally and externally driven processes. When surface modification occurs on a rapid timescale, current surface features give insight on processes shaping the most recent events. History must be inferred from study of structure and stratigraphy underlying current features. Most solar system bodies, including the Moon and asteroids, are chiefly shaped by external processes. The impact-driven resurfacing rate has slowed down historically due to the decrease in size and frequency of the bombarding object population. Although we can certainly observe the effect of later events that have overlapped spatially on earlier events, how much palimpsest-like influence do the larger magnitude earliest events still exert?

Various analytical tools can be used to identify patterns in shape or feature distribution (Clark and Rilee 2010). Spatial harmonic analysis may be performed to identify patterns in the distribution of features to determine the relationship between different regions of the same target body, or for different target bodies. For example, an analysis on the density and angular orientation of ridges and trenches associated with plate tectonics could be used to assess the likelihood of triggering by a single global-scale impact event (Lowman 2002).

P.E. Clark and C. Clark, *Constant-Scale Natural Boundary Mapping*
to Reveal Global and Cosmic Processes, SpringerBriefs in Astronomy,
DOI 10.1007/978-1-4614-7762-4_3, © The Author(s) 2013

3.2   Externally Driven Processes

Erosional/depositional processes, including bombardment, operate on all objects regardless of size and origin, although their surface expressions may well depend on the structure and composition of the surface and interior, as well as the nature and size of the erosional/depositional agent. Every external agent has characteristic signatures (Clark and Rilee 2010), which we will discuss below, meteoritic bombardment being the most common for solar system bodies. Morphological boundaries, such as crater pits, peaks, rims, rings, and rays, result from radial and concentric features formed during large impacts. The very irregular character of the bulk of the asteroid population is an extreme version of cratering, a direct result of bombardment by objects more comparable in size than those bombarding the larger 'regular' planets. Rotationally driven processes may also have fluid dynamical expression observable in the pattern of distribution of surface fluids.

Impact events that are small on the scale of the object will not leave a global signature. On the other hand, encounters with other objects of sufficient energy to affect regions antipodal to the points of shock leave a distinctive global signature, where the 'boundary' is the inner ring of the resulting impact crater. The resulting map would be the equivalent of an equal area (azimuthal equidistant) map. Mega-impacts were most frequent early in our solar system's evolution, their traces partially obscured or erased by later events or processes.

A primary cause of surface modification, erosion and deposition, is dynamically induced movement of fluids. The magnitude and nature of this interaction depend on surface conditions, availability of materials, and driving forces. Fluids may be solids, such as ice; liquids, such as water; or atmosphere fluidized as wind. On the Earth, environmental conditions encompass the 'triple point' of water, which thus may exist in all three phases simultaneously, although one may be predominant. Fluid flow is driven by gravity (toward center of mass), pressure (from high to low), or temperature (affecting dispersal of gases or chemical deposition). Rotational (Coriolis effect) and tidal forces, along with temperature, density, and compositional differences, induce global patterns in mean flow in atmospheres, large liquid bodies, or deformable solids. On the Earth, these are seen as ocean currents, consisting mainly of gyres that flow clockwise in the northern hemisphere and counter-clockwise in the southern hemisphere, with countercurrents at the equator and poles. Terrestrial prevailing wind patterns show a similar pattern in flow with latitudinal zones of tropical trade winds, mid-latitude westerlies, and polar easterlies in each hemisphere. CSNB maps are ideally suited for displaying global patterns and disturbances in these patterns induced by storm systems or climate change as a function of time or environmental conditions using available parameters with patterns in maxima and minima distribution, such as wave height, roughness, or direction; air pressure, density, temperature, or composition.

Fluids induce surface morphology-modifying erosional or depositional processes on solid surfaces (Clark and Rilee 2010). Liquids and deformable solids create networks of channels with distinctive morphologies (e.g., river versus glacial valleys) and a distribution dependent on the underlying structure (e.g., dendritic versus trellis stream distribution). Deposition and erosion may occur in close proximity. Fluid removal and transport of sediment from one area may result in deposition in another nearby due to environmental conditions, including changes in channel depth or width, and encounter with opposing forces that reduce fluid speed, in, for example, glacial outflow and moraine formation, or changes in temperature and composition of the channel bed in, for example, precipitation of limestone-rich sediment. Wind similarly causes erosion and deposition of smaller sediment in, for example, dune formation. CSNB mapping is useful in ascertaining global-scale patterns in erosional networks, potentially associated with underlying tectonically induced structural troughs, such as the Martian rift valleys, as well as depositional networks, such as global distribution of terminal moraines from Earth's last ice age.

CSNB maps illustrating externally driven processes include Figs. 1.5, 4.6, 4.7, 5.1, 5.2, 6.2, 6.4, 6.5, 6.9, 6.10, and 6.11.

3.3 Internally Driven Processes

Internally driven processes form maxima and minima (in elevation or other parameters, such as magnetic or gravitational field amplitude) that result from upwelling or downwelling associated with internal activity cells driving volcano-tectonic (or thermonuclear in the case of stars) activities. Cell formation results from the overturn associated with interior heat transfer through convection or conduction within interior liquid or plasma cells. In planets, the potential sources of heating may be mechanical (tidal deformation) or chemical (interior radioactive nuclides), or both.

Crustal movement, or tectonism, resulting from interior heating, involves crustal deformation or displacement (Clark and Rilee 2010). Tensional movement is associated with crustal thinning, lengthening, block faulting, rift formation, upwelling, and volcanism. Compressional movement is associated with crustal thickening and uplift on the thrust side, shortening, thrust faulting, downwelling, and possibly volcanism. On Earth, crust is generated and consumed in a conveyor-like process, linked with internal convection cells, known as plate tectonics. Crust acts like plates floating on the mantle, generated at tensional, divergent boundaries through upwelling, and consumed at compressional, convergent ones through downwelling. Thus, long volcanic ridges develop at divergent boundaries. At convergent boundaries, uplift on the thrusting plate side, as the compressing crust thickens, is accompanied by long trough formation and heating at the intersection with crust from the other side, resulting in volcanism in the thickened crust. Along plate boundaries, lateral displacement, or transform faulting, also occurs.

Though the style may vary considerably, large volcanic features are easily recognizable (Clark and Rilee 2010). Volcanic shields, domes, or cones typically occur in clusters or chains associated with an underlying heated plume of material. Chains of volcanoes, periodically reactivated, will be associated with plate boundaries in plate tectonics systems, as described above. Further, an isolated chain of progressively younger volcanoes not clearly associated with a plate boundary indicates conveyor-like movement of crust over a mantle hot spot and the potential for a future divergent boundary to form. An isolated cluster of volcanoes with no apparent pattern in age distribution indicates that the crust isn't migrating in relationship to the underlying mantle. Volcanic plains typically occur as low-lying features filling basins that may have formed through external (impact) or internal (tectonic) processes.

CSNB mapping is useful in ascertaining global-scale pattern or lack of pattern in crustal terranes bounded by ridge and trough trees representing the distribution of clusters and chains of volcano-tectonically generated ridges and troughs. Also revealed will be variations in style and timing of volcano-tectonic activity.

CSNB maps illustrating internally-driven surfaces include Figs. 1.3, 1.9, 1.10, 2.4, 2.5, 2.7, 2.8, 4.1, 4.2, 4.3, 4.4, 4.5, 5.3, and 5.5.

3.4 Making Comparisons

CSNB boundaries are identified systematically. The resulting maps reveal global patterns and structure difficult if not impossible to discern in conventional systems, as the next chapters will demonstrate. Keep in mind that merely because a map clearly depicts a plausible theory does not make the theory true.

Volcano-tectonic processes dominate most of the terrestrial planets, and some of the outer planet satellites, but the style and speed of these processes vary considerably. The relationships between volcanic and tectonic features, and the role of externally driven agents, must be ascertained to establish viable formation models, and formulate the relationship between bulk composition, structure, and history of a given body. The evidence for these relationships is provided by comparisons of the number, distribution, shape, and size of bounded units.

That doesn't mean that very large impacts experienced in the past haven't left a signature and influenced subsequent patterns of crustal features. To show this, we made a highly segmented CSNB map bounded by the rim of the largest, most ancient impact structure considered possible, the Gargantuan Basin on the Moon, and looked on the map for evidence of features distributed radially and laterally, revealing thereby the palimpsest-like continuing influence of the early impact.

Just as we may choose to map Earth's oceans contiguously for marine studies or its landmasses for population studies, we may divide any body in a geologically or topographically meaningful manner to suit a particular purpose. CSNB maps provide a uniquely conformal and bandwidth efficient way of doing this. The results may seem surprising at first, perhaps giving the impression of being weakly controlled, but they may illustrate particular themes better than conventional methods. Folding globes of asteroids, for example, may have great educational value and power as a visualization method. As Stooke (1998) and Berthoud (2005) have shown, novel methods can produce meaningful maps of planetary bodies or celestial objects poorly suited to traditional cartography. CSNB maps appear to be an optimal way to portray and demonstrate genetic relationships between structures as global systems.

Chapter 4
Mapping the Earth

4.1 Earth's Dynamic Context

A major earthquake in China, a cyclone in Burma, a volcanic eruption in Chile; and that's just one week's havoc. Ongoing crises include melting Himalayan glaciers, persistent North American drought, Antarctic ice sliding into the sea, disappearing bees, and blooming algae. We have the capability, through remote sensing and data processing technology, to monitor such acute and chronic disasters in great detail. Is there a way of managing and interpreting such overwhelming detail in order to obtain global-scale context, reveal patterns, improve predictions, and mitigate problems? We would argue that the CSNB mapping technique, described in detail in the introductory chapters, is a tool that could provide such insight. Our approach assumes that nothing can happen 'here' without affecting what is happening 'there' on the Earth's surface, whether we are talking about plate tectonics or major storm systems. This is not to denigrate but to supplement existing forecasting models with a tool to focus the global component. In CSNB, any set of identifiable boundaries can be selected as critical and used to define map edges. Thus, any hypothesis, based on the set of critical boundaries it predicts, may be evaluated and tested based on the shapes and relationships illustrated on resulting maps. We use examples from terrestrial tectonics and global current patterns.

Earth's hemispherical asymmetries are many: *Water* makes a circle-current in the south yet bisects the north. Felsic continental and mafic oceanic *Crust* are distributed with a north/south asymmetry. *Ice* is a concentrated crustal load at south, and a distributed (floating) load north. *Air* mirrors the water: north a circular jet stream dominates while south, vapors collect at a point, the Antarctic High.

Some other striking features appear in the details. The unexpected hemispherical reach of the American Cordillera is perpendicular to, and commensurate in size to the Equatorial East Pacific Current, the ocean's primary 'returning current' There is current sufficient to nurture coral, when stagnancy ought to rule in the Gulf of Mexico, Caribbean, Red Sea, and Persian Gulf. While the Arctic's through-flow is

P.E. Clark and C. Clark, *Constant-Scale Natural Boundary Mapping*
to Reveal Global and Cosmic Processes, SpringerBriefs in Astronomy,
DOI 10.1007/978-1-4614-7762-4_4, © The Author(s) 2013

minimal, Antarctica's two great gulfs are each a source of powerful basement currents moving away from the (rotation induced) Southern Current; a balancing act worthy of a fluid gyroscope.

4.2   Tectonic Activity

Early in the development of the paradigm of plate tectonics, Wilson and others hypothesized that the next, and ongoing, episodes of plate tectonics would involve repeated ongoing 'Wilson cycles' of ocean expansion and continent spreading in one hemisphere, accompanied by ocean contraction and continent collision in the other. As new oceanic crust is generated at rifts called mid-oceanic ridges, sea-floor spreading and ocean expansion is driving continents apart in the Atlantic; meanwhile, subduction of old, dense oceanic crust is accompanied by volcanic activity and mountain formation in continental crust in trenches along the Pacific rim, shrinking the Pacific at the expanse of the Atlantic. After the Pacific closes and plates collide, reactivated rifts will drive continents apart and expand the Pacific, and oceanic crust will be subducted along the rim of the Atlantic while the Atlantic shrinks.

Evidence cited in support of the Wilson cycle includes observations that supported continental drift: clear geological relationships between opposite and adjacent portions of continents on either side of spreading ridges and the creation of volcanic island chains as plates apparently migrate over a hotspot. Compelling support is provided by observation of new seafloor erupting with magnetic alignment patterns symmetrical on both sides of mid-oceanic ridge and apparently 'pushing' the continents on either side away from each other.

Present observations do not support the initial simplistic model of subduction, which is obviously not well understood, and thus may call the Wilson cycle itself into question. What is the driving force behind plate tectonics? What is the real nature of plate boundaries and movement? How do plates and other tectonic features relate to spreading ridges and subduction trenches? Can we make global CSNB maps illustrating the Wilson cycle's basic tenets?

We construct maps bounded by spreading centers in the Atlantic/Indian Oceans or the Pacific Ocean using either topography (Fig. 4.1) or geology data (Fig. 4.2) as the data field. Note the similar pattern in shape and distribution of oceanic crust and trenches about mid-oceanic ridges. However, the Pacific ridge-bound map exhibits additional complexity in the map interior, with more prominent, semi-continuous relationships of mid-oceanic ridges that bear some resemblance to small incipient rifts.

The waterline test discussed in Sect. 2.6 could be applied to maps with these 'critical' tectonic boundaries to see if stress interference patterns correspond to significant physical features, and thereby validate the significance of the chosen boundaries as drivers of long-term trends in tectonic activity.

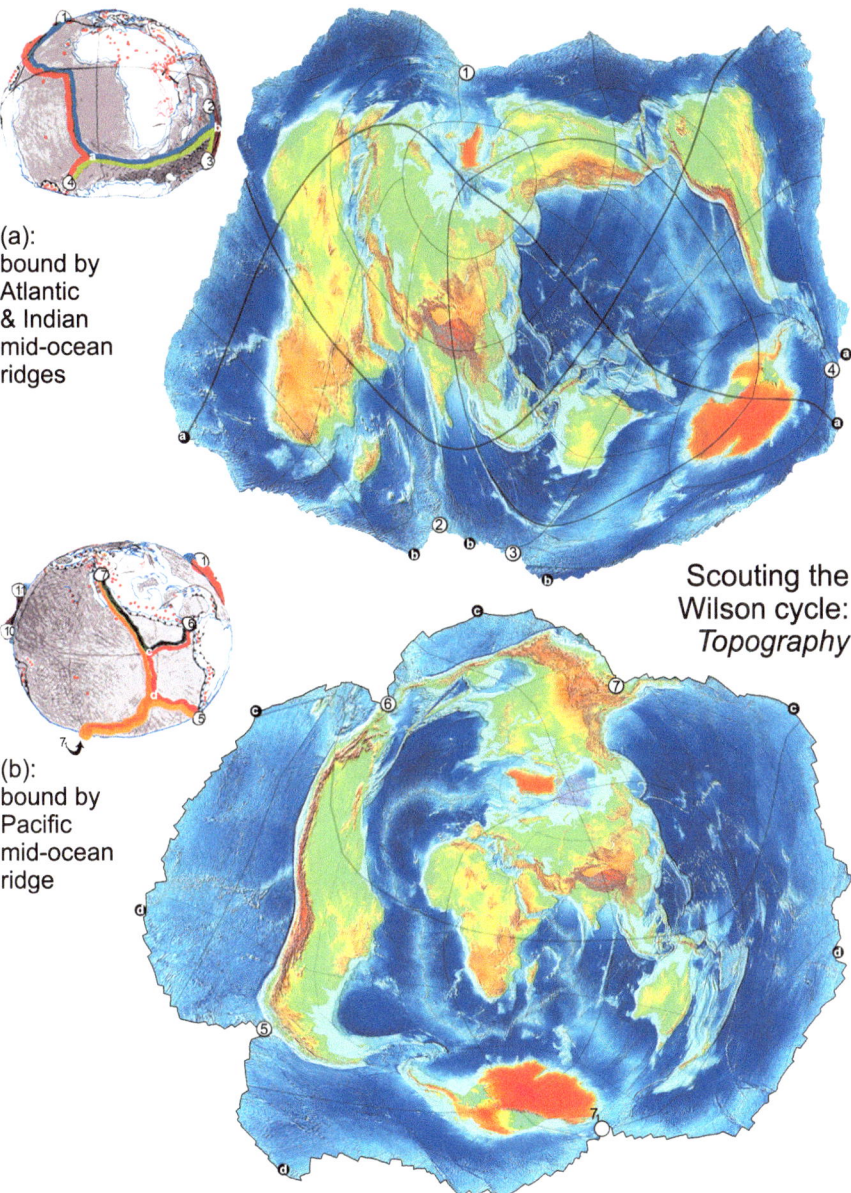

(a):
bound by
Atlantic
& Indian
mid-ocean
ridges

Scouting the
Wilson cycle:
Topography

(b):
bound by
Pacific
mid-ocean
ridge

Fig. 4.1 CSNB topography maps bounded by (**a**) slow mid-oceanic ridges suggested by Dave McAdoo and (**b**) fast mid-oceanic ridges suggested by Paul Lowman, for comparison as discussed in Sect. 4.2 (Source data MGG images, NOAA)

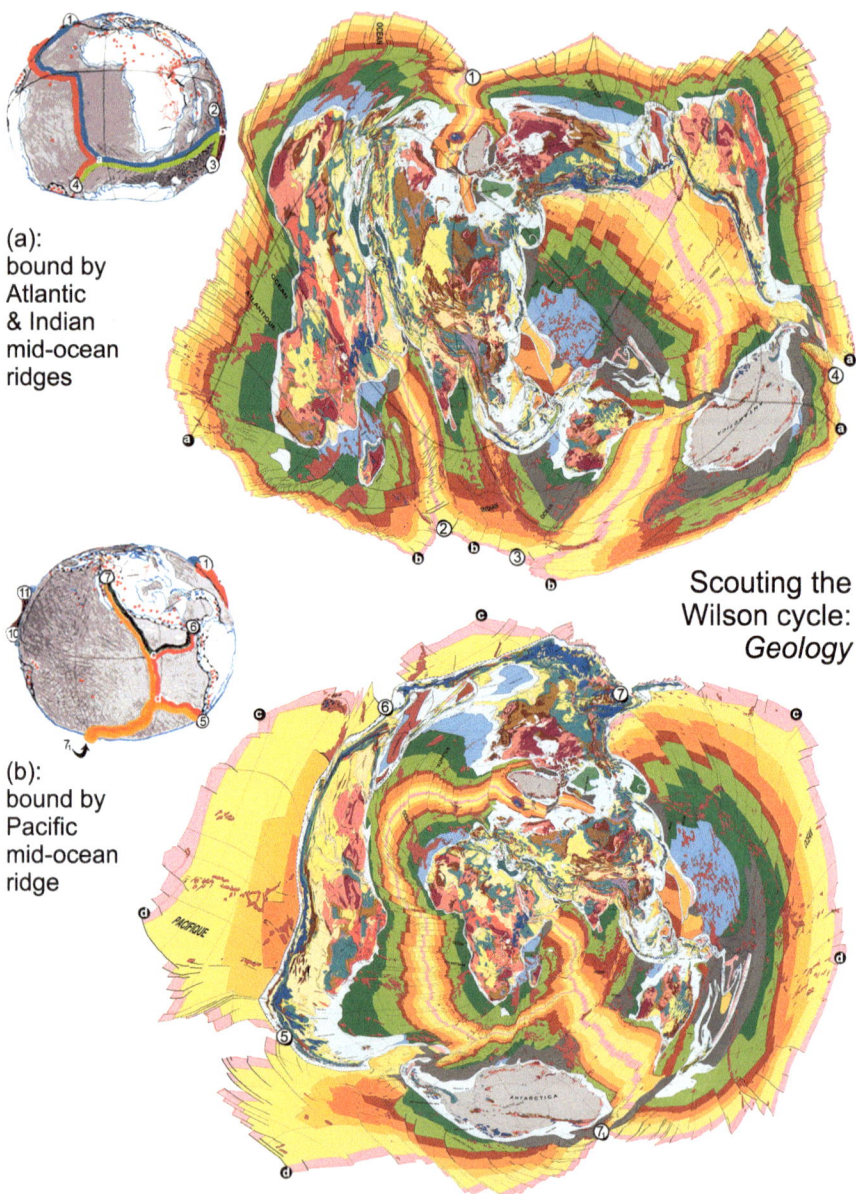

Fig. 4.2 CSNB geology unit maps bounded by (**a**) slow mid-oceanic ridges suggested by Dave McAdoo, and (**b**) fast mid-ocean ridges suggested by Paul Lowman, for comparison as discussed in Sect. 4.2 (Source data Geological Map of the World at 1:250000000, 2nd Edition. Philippe Bouysse et al. © CCGM/CGMW 2000)

For broader panoramas, we construct maps with other critical boundaries, including the Wallace line in Indonesia, which links tectonics and biological evidence (Whitmore 1981). The line, named after its discoverer, nineteenth-century naturalist Alfred Russel Wallace, marks a sharp dichotomy in distribution of fauna, separating the lower half of the Indonesian island arc (and Australia) from the upper half, potentially due to earlier separation of an Australian plate (Fig. 4.3). The map highlights the isolation of Australia and surrounding islands from Asia. We derive an additional map that has as its edge all tectonic boundaries, including rifts and spreading centers, except mid-oceanic ridge spreading centers; this map depicts the mid-oceanic ridge, Earth's largest physical feature, as a whole (Fig. 4.3). This maps shows an east/west dichotomy in the distribution of oceanic and continental crust.

Finally, at the most segmented extreme of CSNB mapping, we produce a map (Fig. 4.4) that has as its edge all global tectonic activity of the last one million years (Lowman 2002). A version of this highly interrupted map is designed to be cut out, with tabs to connect and fold it up into a 3D mechanical model with flexible, 'active' plate boundaries, capable of mimicking tectonic activity. Such a model, perhaps in conjunction with the map and models shown in Fig. 2.7, could reveal insights into the postulated subduction of the proposed Farallon plate.

4.3 Watersheds, Watercourses, and Weather

We now use watersheds, or ridge trees, of Earth's two major landmasses as critical boundaries (Fig. 4.5) to produce a pair of complementary maps. In the first case, the ridge systems of Eurasia/Africa form the critical boundary, and North/South America is a watercourse; in the second case, the ridge systems of North and South America form the critical boundary, and Eurasia/Africa is the watercourse. Clearly, we could use a more extensive (branched) or less extensive (pruned) system of ridgelines, and produce a map edge that is more or less complex, respectively. Arrayed at the map's peripheries are the various watersheds of interrupted landmasses, adjacent to the bays, gulfs and seas into which they debouch. Inland basins are attached at their lowest saddle; relative sizes and absolute shapes of circumferential basins may be read directly. The ridge tree forming each map's edge is seen in the complementary map as a connected whole. Note that Earth's continuous watery surface is seen as a unified shape, within which the several oceans (Pacific, Atlantic, Indian, Southern and Arctic) are depicted in proper relative proportions.

Watercourses form an interesting backdrop, and provide a much larger meteorological display arena than is conventionally used, for observing the movement of weather systems and any systematic trends in weather relative to these two major landmasses. In each map, the northern jet stream could be plotted in its actual circular pattern. How effectively does the interaction between the jet stream and the topography, the secondary ridgeline system within the basin, control weather patterns?

As an example, we provide a series of snapshots of global cloud patterns during a three-week, three-storm interval of the 2009 hurricane season (Fig. 4.6). Global

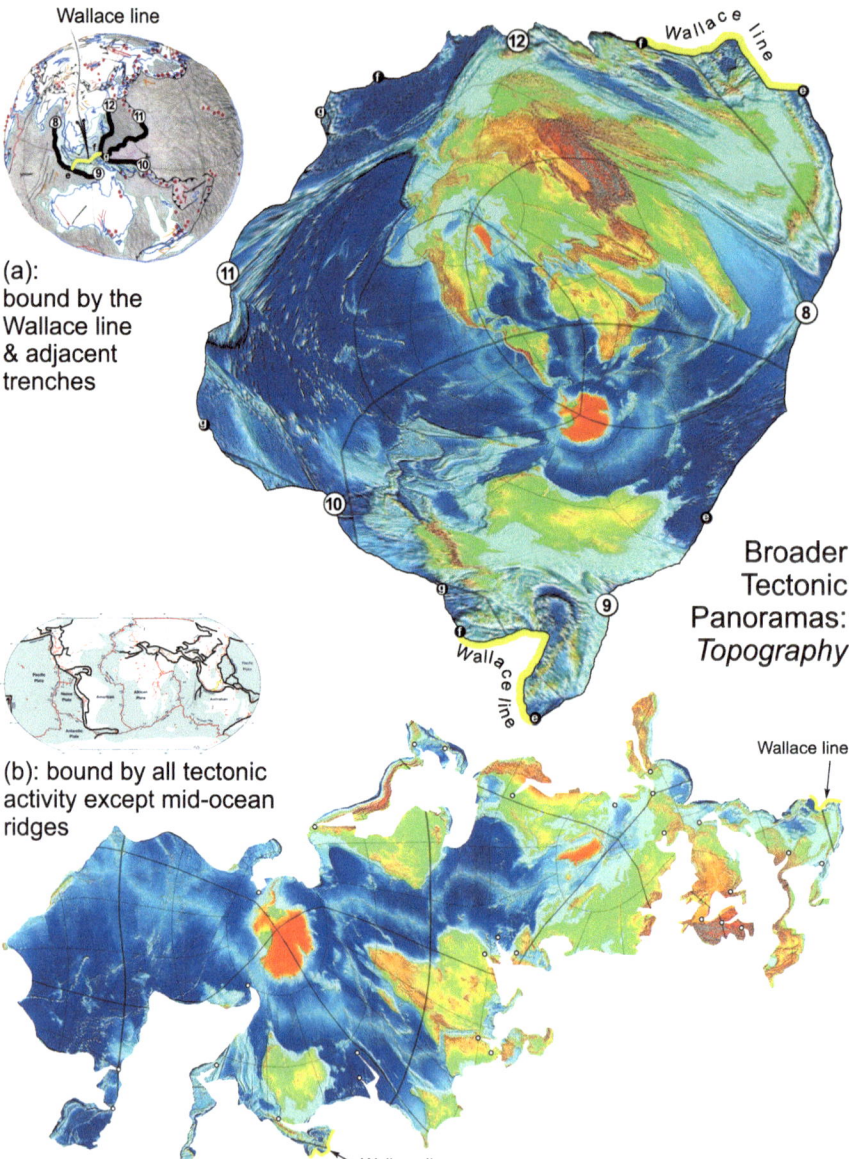

Fig. 4.3 CSNB maps with (**a**) a tightly focused multidisciplinary boundary, and (**b**) a wide-angle multidisciplinary boundary chosen to feature the mid-ocean ridge (Source data MGG images, NOAA. (**b**) inset adapted from NASA/GSFC, Global Tectonic Activity Map, Paul Lowman)

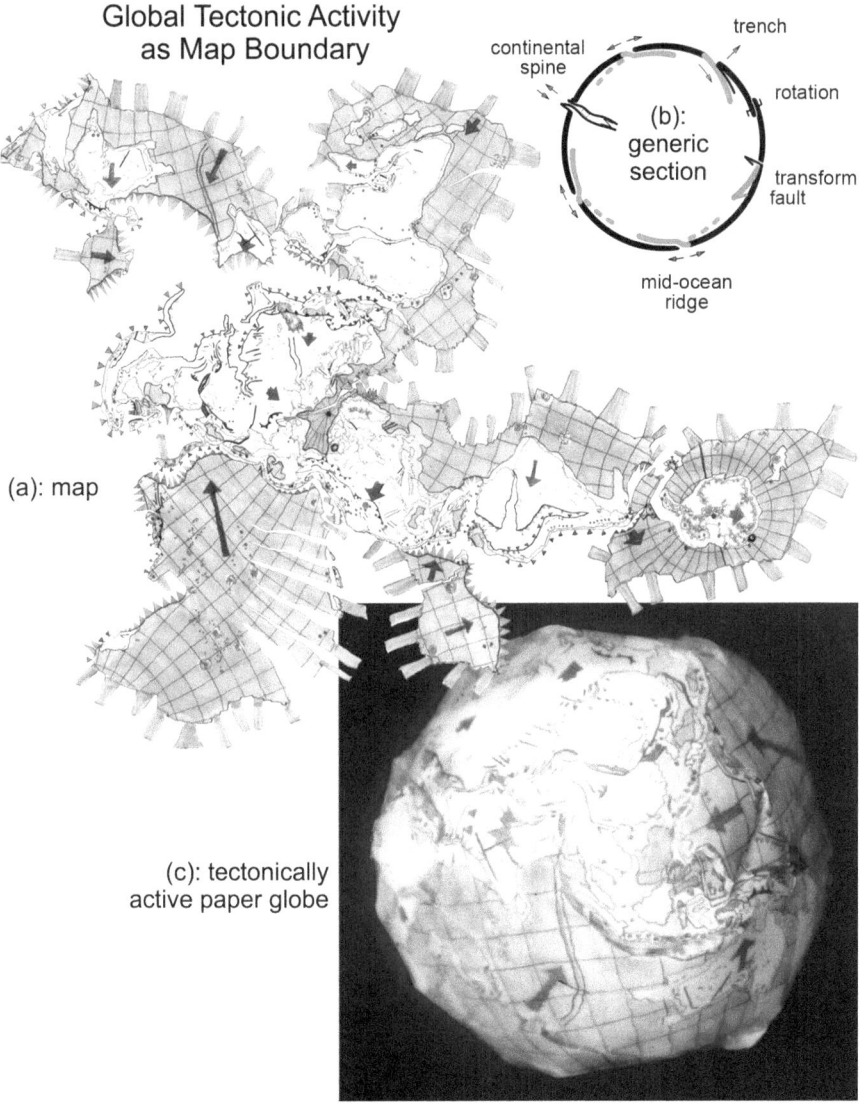

Global Tectonic Activity as Map Boundary

(a): map

(b): generic section

continental spine

trench

rotation

transform fault

mid-ocean ridge

(c): tectonically active paper globe

Fig. 4.4 CSNB tectonic plate map with (**a**) all plate boundaries, and with (**b**) tabs, for making 'tectonically active' 3D model (Clark 2003) (Source data courtesy of NASA/GSFC, Global Tectonic Activity Map, Paul Lowman)

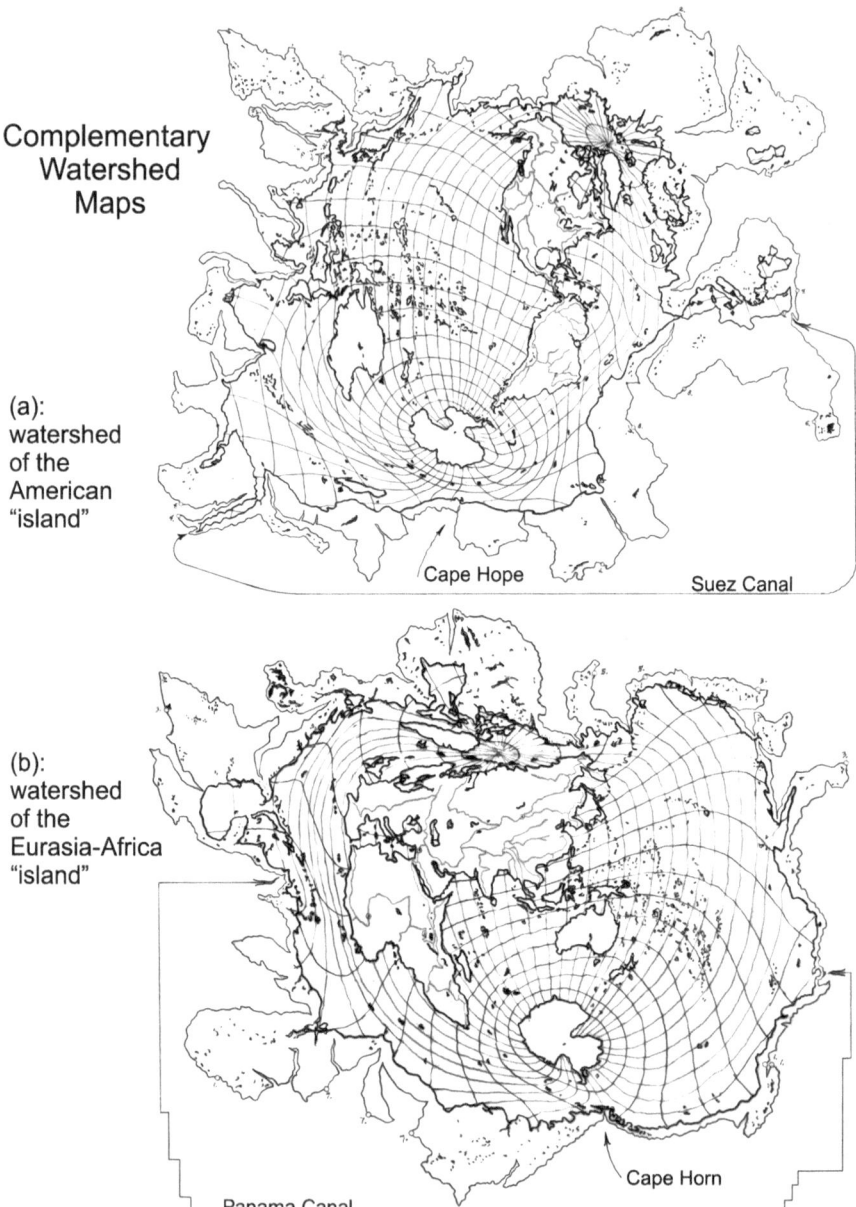

Complementary
Watershed
Maps

(a):
watershed
of the
American
"island"

Cape Hope   Suez Canal

(b):
watershed
of the
Eurasia-Africa
"island"

Cape Horn

Panama Canal

Fig. 4.5 Watersheds, as defined by ridge trees, for (**a**) Eurasia/Africa and (**b**) North/South America, as described in Sect. 4.3. C.S. Clark, unpremiated entry, 27th ACSM Annual Map Design Competition, 2001

Fig. 4.6 Animation stills showing a global sequence of major storm cloud cover (Prepared with map shown in Fig. 4.5a (Source data Earthdesk, www.xericdesign.com))

patterns, such as El Nino and La Nina, suggest a relationship between atmospheric pressure and oceanic temperature in one hemisphere and major storm production in the other. Such maps, animating major storm systems seasonally during the course of a decade, would be ideal for confirming these relationships.

4.4 Ocean Currents

Prevailing winds are horizontal drivers, and density differences vertical drivers for major surface and subsurface currents respectively. What is the relationship between major ocean currents on a global scale? How are these related to weather systems or climate change? Figure 4.7 shows maps of surface, mid-level, basement, and seasonally reversible currents in azimuthal equidistant and CSNB projections. The boundary of the CSNB map consists of Antarctic ridges, minimizing distortion and providing a recognizable connection between current systems and the two large Antarctic bays. The most extensive cold, deep currents apparently originate in and dominate the south circumpolar region. Cold currents sweep by the west coasts of the landmasses, and warm currents by the east coasts of landmasses. Deeper currents dominate in the Pacific hemisphere, and shallower ones in the Atlantic.

The Antarctic High is well studied in the map bordered by valleys of Antarctica. Katabatic winds flow downward as a function of a valley's size and relative proportions (things perfectly captured by CSNB maps), and go 'inward'. Given that context, a Greenland High–edged CSNB map, with ridges formed by Greenland's peaks, would be an excellent northern hemisphere meteorological format.

Ocean currents internal to the Pacific, Atlantic or Indian Oceans could be shown as map-edge. And we may pursue this theme: currents internal to these basins could remain zipped, but Earth's watery surface interrupted only at the Bering Sea and the Straits of Indonesia, and land interrupted along *both* of the ridge trees used in Fig. 4.5. Such a map would show major oceans as individual units.

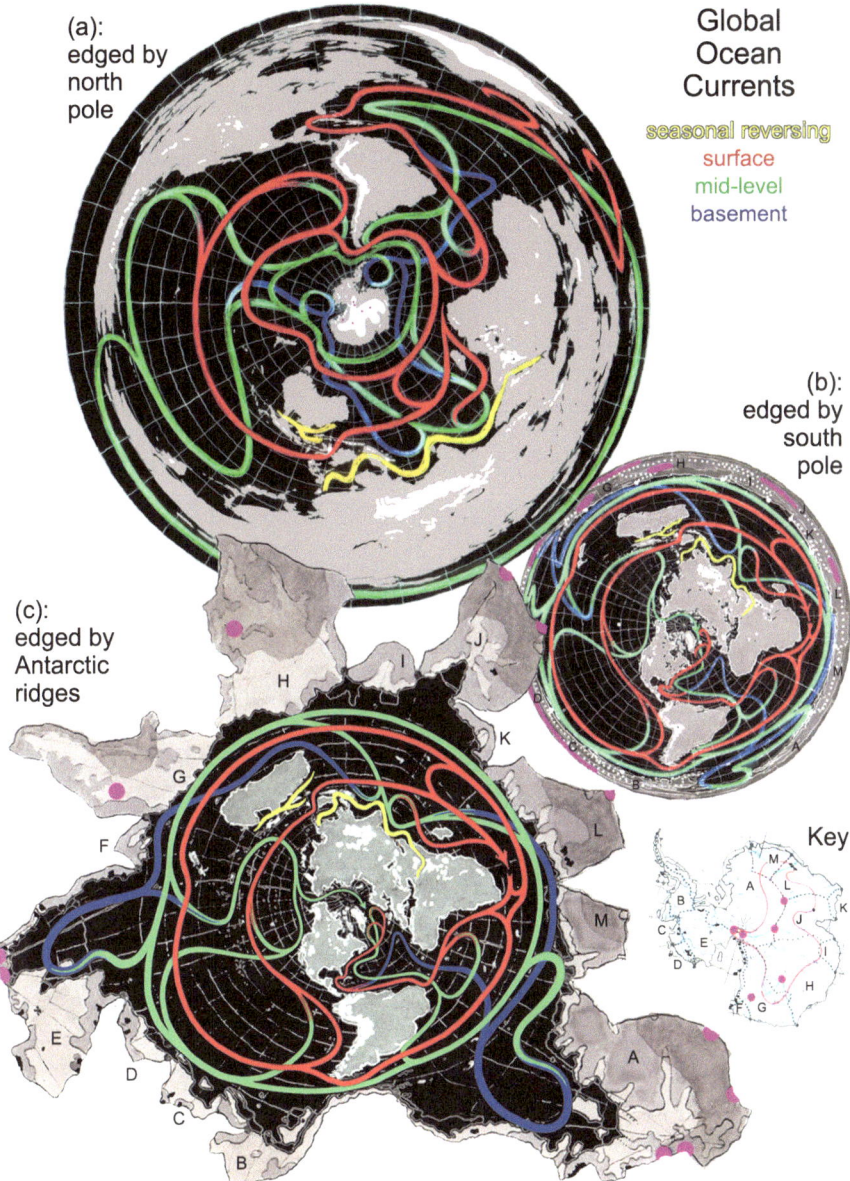

Fig. 4.7 Currents in azimuthal equidistant map edged by (**a**) North Pole and (**b**) South Pole, and (**c**) a CSNB map bound by Antarctic ridges (Clark 2003) (Source data Ackerman 2000)

Chapter 5
CSNB Mapping Applied to Regular Bodies

5.1 Overview of Application

In this chapter, we apply CSNB technique to regular solar system bodies to illustrate the nature of processes prevailing on these bodies. These oblate spheroids include the Moon, clearly dominated by external bombardment, Venus, apparently dominated by volcanism, and Mars, where externally and internally driven processes compete and have dominated at different times in the planet's history. We also outline our plans to apply this technique to Mercury, and to icy and outer solar system bodies.

CSNB has now been used to produce global maps of bodies lying on a continuum between externally and internally driven surface morphology. For the Earth and Venus, representing the internally driven end of the continuum, ridge and trough boundaries are apparently driven by internal activity cells; thus CSNB maps show large-scale patterns of, for example, gravity and magnetic anomalies. For asteroids, representing the externally driven end of the continuum, impact-generated irregular facet edges become critical boundaries; highly segmented maps derived from these bodies provide insight into bombardment history, as discussed further in Chap. 6. Although we typically see the Moon mapped in a conventional projection, it is far more asteroid-like in terms of surface modification. Here, we appropriately map our Moon in highly segmented fashion, with impact-linked global features as critical boundaries, to reveal the continuing influence of old, large impacts. If Mars, a body with competing surface modification processes, were also similarly mapped, the result might reveal the palimpsest-like influence of its proposed equivalent gargantuan basin (Wilhelms and Squyres 1984).

P.E. Clark and C. Clark, *Constant-Scale Natural Boundary Mapping to Reveal Global and Cosmic Processes*, SpringerBriefs in Astronomy, DOI 10.1007/978-1-4614-7762-4_5, © The Author(s) 2013

5.2 The Moon

The lunar surface has been impact-dominated for most of its history; thus, we anticipate that global-scale morphological features represent radial and concentric features associated with the biggest impact events. After initial crust formation, impact
events triggered volcanism in the largest impact-generated basins. Volcano-tectonic
activity was largely, although not entirely, confined to the basins. The impact-driven
activity rate has slowed historically due to the decrease in bombarding projectile
size and frequency. Although we can certainly observe the effect of later events
overlapping spatially on earlier events, how much palimpsest-like influence do the
large-magnitude earliest events still exert? Has the earliest network of global-scale
radial and concentric fractures and ridges been reactivated by later impacts and if so
to what extent?

A number of workers (Cadogan 1974; Whitaker 1981; Wilhelms 1982, 1987)
have proposed that much of the underlying structure of the nearside was shaped by
a major impact event. The resulting impact feature, known as Gargantuan Basin,
can still be seen as much of the continuous outline of western nearside Oceanus
Procellarum, as well as in the discontinuous mare and terra features distributed
concentrically and radially to the original basin. The interior of such a basin would
be expected to have a thinner crust, which could explain the globally systematic
changes (as a function of radial distance to impact center) in composition and age
of the basalts (e.g., KREEP or Ti abundance) that subsequently erupted (Clark
1985; Clark and McFadden 2000).

We have applied the CSNB approach to the Moon using the earliest multi-ring
basin maps (Wilhelms 1987) superimposed on airbrush shaded relief and surface
markings map (USGS 1980, 1981, 1992) (Fig. 5.1) and on LOLA topography
(Kreslavsky and Head 2012) (Fig. 5.2). Gargantuan Basin is the center of the CSNB
map and South Pole–Aitken Basin, roughly antipodal to Gargantuan, forms the segment tips. Critical boundaries, radial and concentric to Gargantuan, are features
with both positive relief (wrinkle ridges, escarpments, mountain chains) and negative relief (rifts, rilles, valles). Often these features are associated with basin rings
and radials associated with later impacts that apparently reactivated earlier faults
and fractures. This trend is particularly clear among older materials on the farside,
while the thinner-crust nearside is dominated by more recent flood basalts. The
CSNB map also clearly shows the global distribution of Nectarian basin and Imbrian
basin impact deposits.

5.3 Venus

Like Earth, Venus has a relatively young surface consisting of two major terranes,
both of volcano-tectonic origin. But unlike Earth, Venus' surface is disproportion-
ately covered by mafic volcanic rocks: huge basaltic lava flows, so that flood basalts
extend hundreds of kilometers. Impact features on Venus are far more prevalent

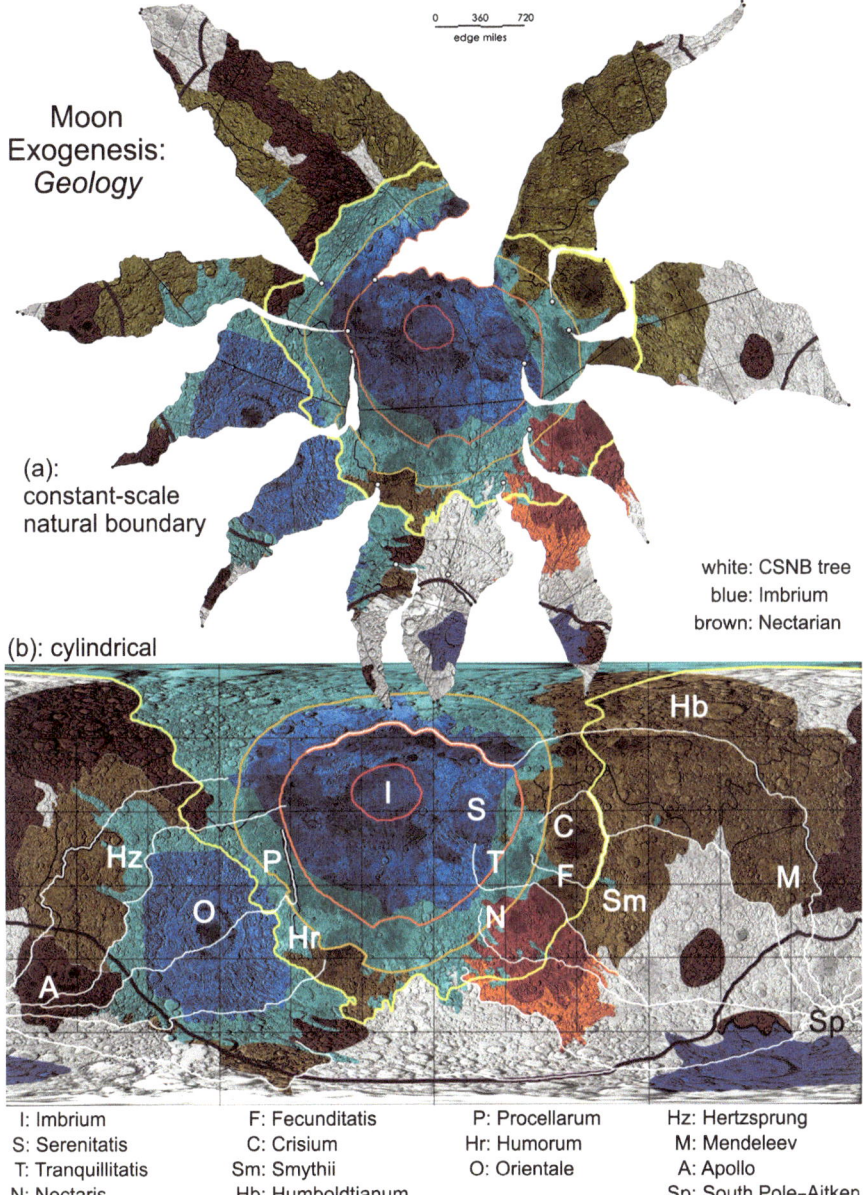

Moon
Exogenesis:
Geology

0   360   720
edge miles

(a):
constant-scale
natural boundary

(b): cylindrical

white: CSNB tree
blue: Imbrium
brown: Nectarian

I: Imbrium	F: Fecunditatis	P: Procellarum	Hz: Hertzsprung
S: Serenitatis	C: Crisium	Hr: Humorum	M: Mendeleev
T: Tranquillitatis	Sm: Smythii	O: Orientale	A: Apollo
N: Nectaris	Hb: Humboldtianum		Sp: South Pole–Aitken

Fig. 5.1 Lunar (**a**) CSNB and (**b**) conventional maps showing geological units and critical boundaries formed by concentric and radial features resulting from oldest putative mega-impact (Clark
and Clark 2006a) (Source geological units Wilhelms (1987) and map background USGS (USGS
1980, 1981, 1992) courtesy of USGS)

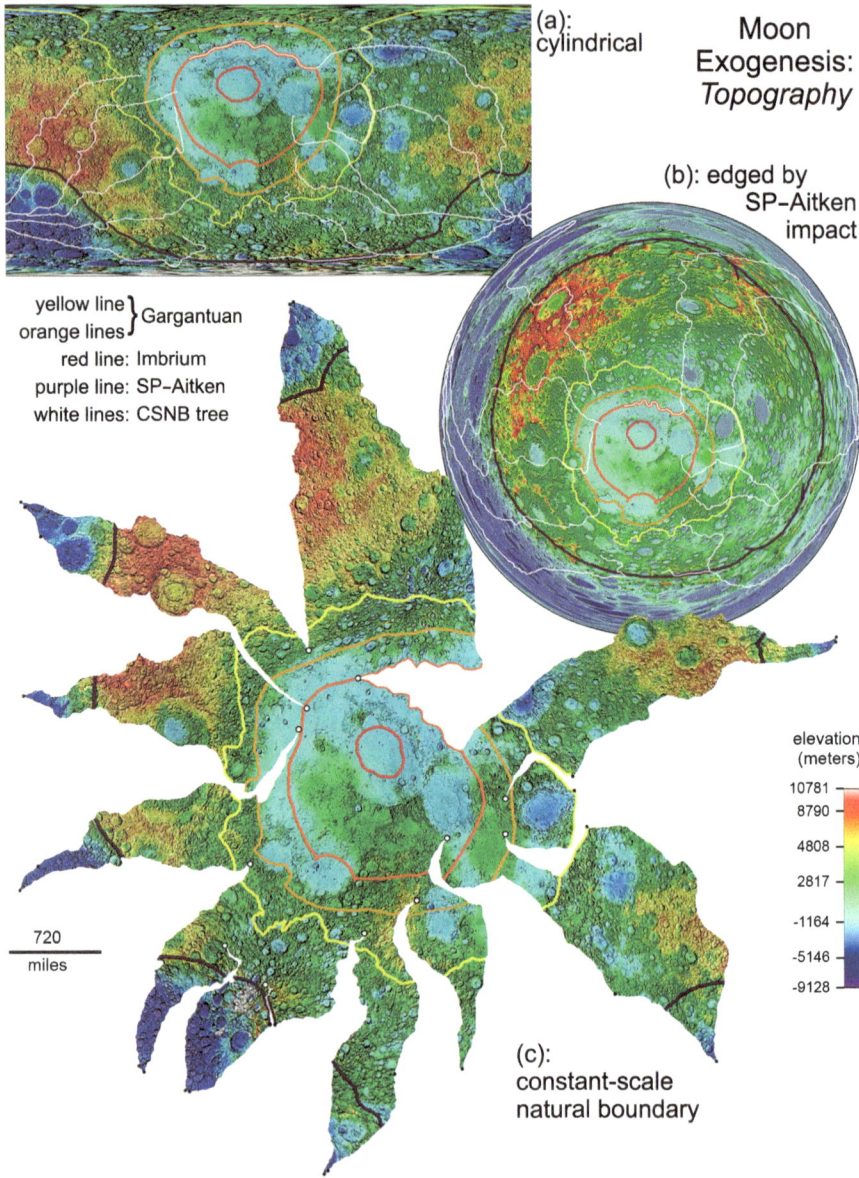

Fig. 5.2 Lunar (**a** and **b**) conventional maps and (**c**) CSNB topography maps where tips of CSNB map segments form the center of South Pole-Aitken Basin (Clark and Clark 2006b). Source data LOLA topography (Kreslavsky and Head 2012), Courtesy of NASA (Source data for map (**a**) and (**b**) Courtesy of USGS (Trent Hare))

than on Earth, indicating that Venus has not been resurfaced as recently as Earth, and is currently less geologically active. Venus has fewer craters than the Moon, Mars, or Mercury, all of which have far less extensive atmospheres. Its largest craters are randomly distributed.

When combined, these observations indicate that Venus resurfacing resulted from the latest episode of global volcanic activity rather than plate tectonics, which would not produce the surfaces of the uniform age implied by the random crater distribution. This suggests that at least periodically Venus had resurfacing rates comparable to the Earth's.

Thus, Venus is a planet clearly dominated by internally driven processes generating sinuous ridges and valleys, from which we identify critical boundary trees using Magellan topography data (McNamee et al. 1993). Figure 5.3 (Clark and Clark 2007) shows these networks in CSNB in contrast to a conventional projection. Note that the ridge tree does not include Venus' highest peak, Maxwell Montes, because it is an outlier (Maxwell 1870).

What insight can CNSB technique provide when applied to massively resurfaced and volcanic Venus? On the valley-bound map (bottom) (Fig. 5.3b), regions considered oldest based on tesserae occurrence (Alpha Regio, Tellus Regio, eastern Ishtar and western Aphrodite Terrae) appear as isolated archipelagoes, inter-connected by equatorial highlands. The major ridge swarms found in the lowest-lying regions on Venus lie perpendicular to the boundaries at the edge of the map. On the ridge-bound map (top) (Fig. 5.3a), the edge is the string of equatorial rift zones. Here, the north/south hemispheric dichotomy is emphasized (right/left as the figure is oriented), with the southern hemisphere being distinctively subdued relative to the northern hemisphere.

5.4 Mars

The global-scale surface features on Mars manifest both internally driven volcano-tectonic and externally driven bombardment origin. Mars has an obvious hemispheric dichotomy: volcanic complexes intruding into minimally cratered lowlands in the north versus, in the south, uplands characterized by heavily cratered terrane and two major impact basins, including Hellas. This implies a crustal dichotomy and internally driven processes, as on Earth and other terrestrial planets. Dohm and coworkers (2001, 2005) suggest a 3000-km mega-impact in Arabia gave rise to Tharsis volcanism. Frey and coworkers (2002) have called the Mars crustal dichotomy "the fundamental unsolved problem in Martian crustal evolution." This bathtub ring–like transition zone is a diffuse, relatively narrow, severe (several kilometer) vertical transition (Mutch and Saunders 1976; Smith et al. 2001). Does this feature represent more than a topographic boundary? Is it reflected in other global datasets? These are questions CSNB is designed to answer.

Even a conventional cartographic map (Fig. 5.4a) clearly indicates the north/south dichotomy. The northern plains act as a global sink, and drainage roughly

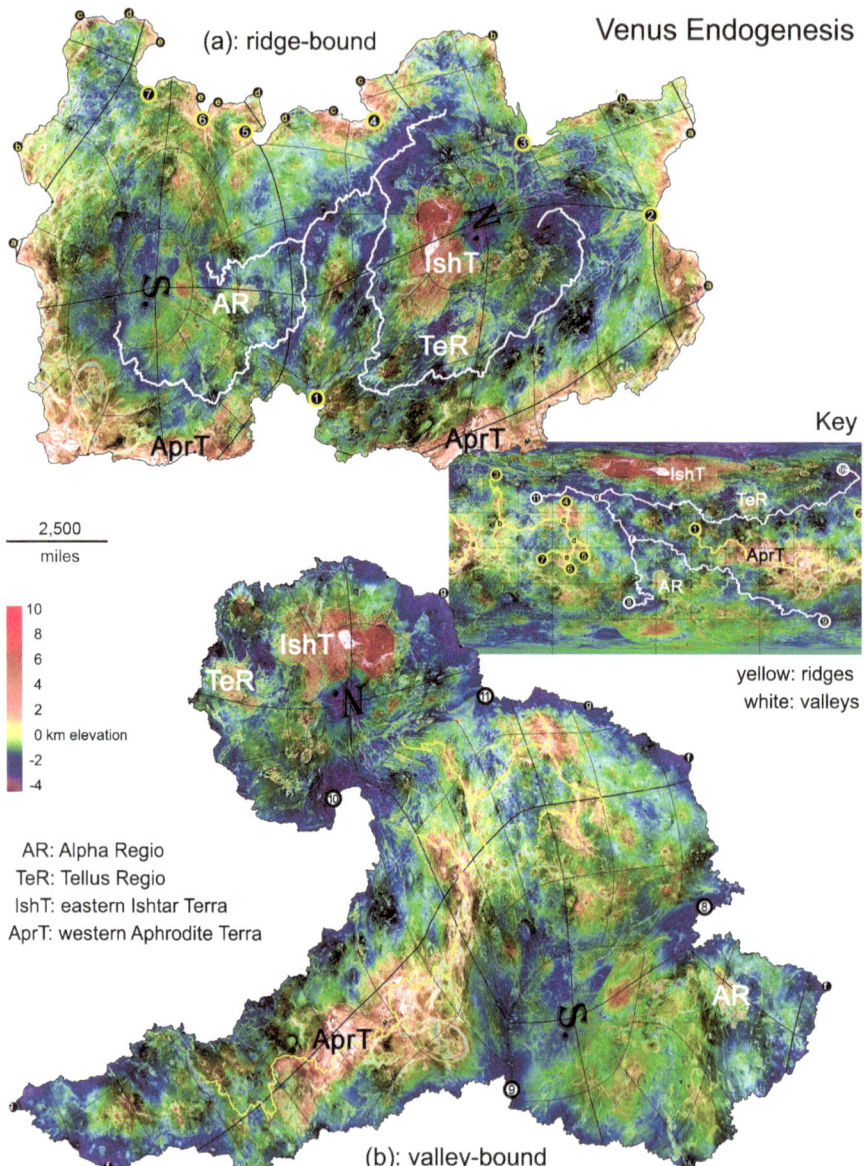

Fig. 5.3 Venus tectonically driven (**a**) ridge-bound and (**b**) valley-bound maps (Clark and Clark 2007) based on Magellan topography (McNamee et al. 1993), as discussed in Sect. 5.3 (Source data map and key Magellan, Courtesy of NASA and USGS)

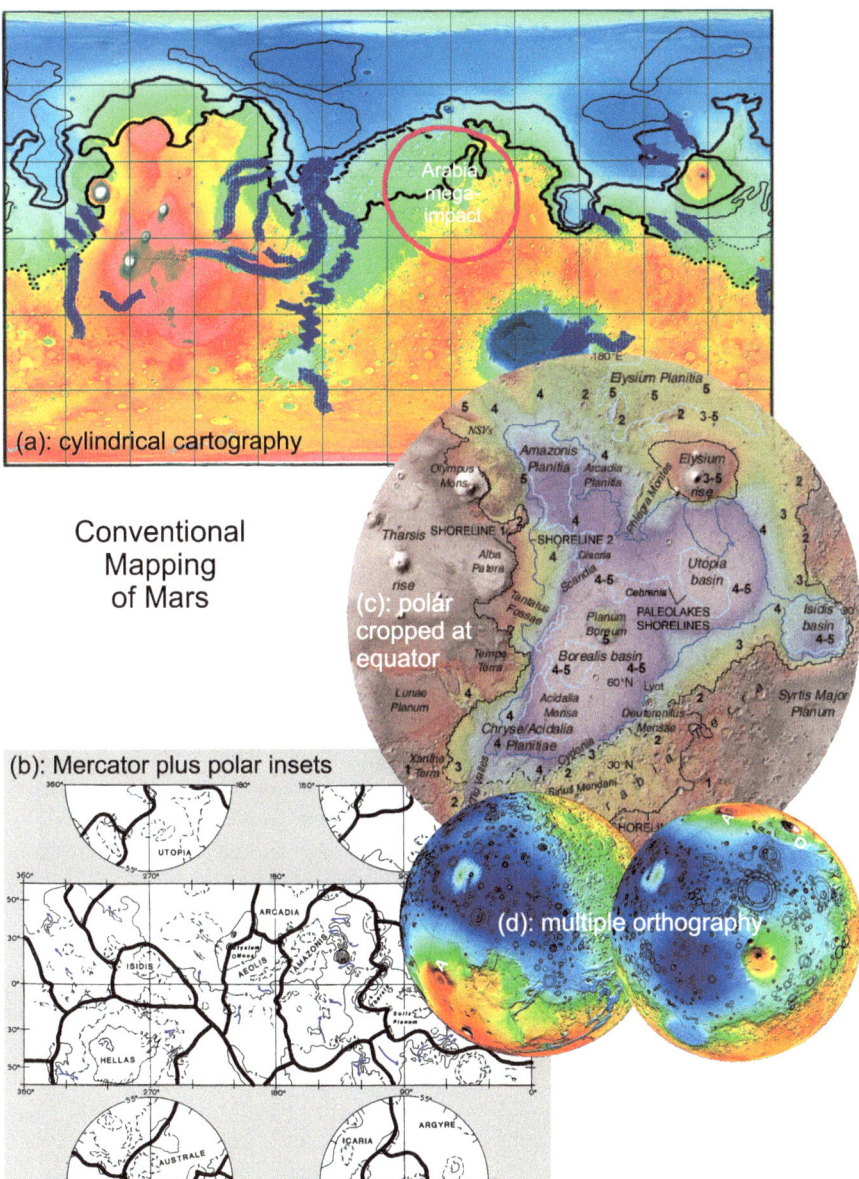

Fig. 5.4 Mars conventional map projections. (**a**) MOLA data map (Courtesy of NASA and USGS) North/south dichotomy distorted (Smith et al. 2001), (**b**) geology drawing of interrupted basins (De Hon 1995), (**c**) geology map of cropped lowlands adapted from Fig. 2 in Fairén et al. (2003), © 2003 Elsevier Inc., and (**d**) quasi-circular depressions seen from limited points of view adapted from Fig. 2 in Frey et al. (2002), © 2002 by the American Geophysical Union

aligns with the map's longitudinal edges. However, the map's straight sides interrupt the continuity in terranes and its top and bottom edges clearly distort the polar regions. Poles are often mapped separately, which restores conformality but disrupts polar–mid-latitude relationships. This is frustrating because the scale of the dichotomy demands global appraisal, at least for matters of slope, e.g., dune migration, water or lava flow. De Hon (1995) opportunistically joins polar insets to a cropped Mercator, with insets partitioned hemispherically near basin divides, to aid the eye in leapfrogging interruptions (Fig. 5.4b). Because polar shape is preserved, we may follow the global gradient with fragmentary clarity. Fairén and coworkers (2003) use another approach to illustrate Mars watershed: a north pole–centered equal azimuth map cropped at the equator (Fig. 5.4c). A companion southern hemisphere map would show excluded uplands, giving a global picture in two conformal halves. However, on such maps, distortion increases as a function of distance from the pole. Frey and coworkers (2002) map the distribution of quasi-circular depressions in two as-the-eye-would-see-it views (Fig. 5.4d). Peripheries are seen obliquely, thus, many views are needed to appraise pattern and structure, and such appraisals will be biased to local perspectives. All three approaches rely on the mind to provide the completing synthesis.

Mars exhibits a rough northern/southern hemisphere dichotomy, with a roughly equatorial 'dichotomy zone'; thus, we generate two maps (Fig. 5.5) (Clark 2007). The ridge-bound map separates continental watersheds above the rift zone; the tree is drawn between maxima through saddles downhill along ridgelines. The trough-bound map separates continental watercourses below the rift zone; the tree is drawn between minima through saddles uphill along channels. Ridges divide the planet into districts called dales, and troughs, or channels, divide the planet into districts called hills. Note the unusual situation evident in the deep equatorial rift zone of Valles Marineris: unlike elsewhere on Mars, the ridge and channel lines are adjacent. As might be expected from erosional geomorphology of uplands and lowlands, the ancient southern highlands are well branched with ridges.

The highest ridges and lowest channels, or valley-lines, were found by visual inspection of lines of slope using multiple sources and checked against other data. Valley-lines per se have not been previously reported, but we find that Arcadia Planitia (cutting an Elysium peninsula) is 500 km west of the location implied by Fairén and coworkers (2003). We find a bar between Icaria and Amazonis basins in Ionia Terra, 700 km northwest of where others have located it (Banerdt and Vidal 2001; De Hon 1995). Note that the critical trees do not include Mars' highest and lowest points: Olympus Mons and Hellas basin. These points are discontinuous outliers, skirted by boundaries needed for framing Mars as a global system (De Hon 2010). Geomorphologically speaking, critical points forming global-scale ridges or valleys do not typically include an object's most extreme peak or pit (Morse 1959).

As map background, we used MOLA topography (Smith et al. 2001). We plotted putative shorelines (Parker et al., 2002) using landforms and graticle intersections as benchmarks. We added shorelines south of the equator at a steady contour, a simplistic assumption given regionally complex geomorphology, but a starting point nonetheless (Fig. 5.5a). Present-day topography undoubtedly deviates from ancient

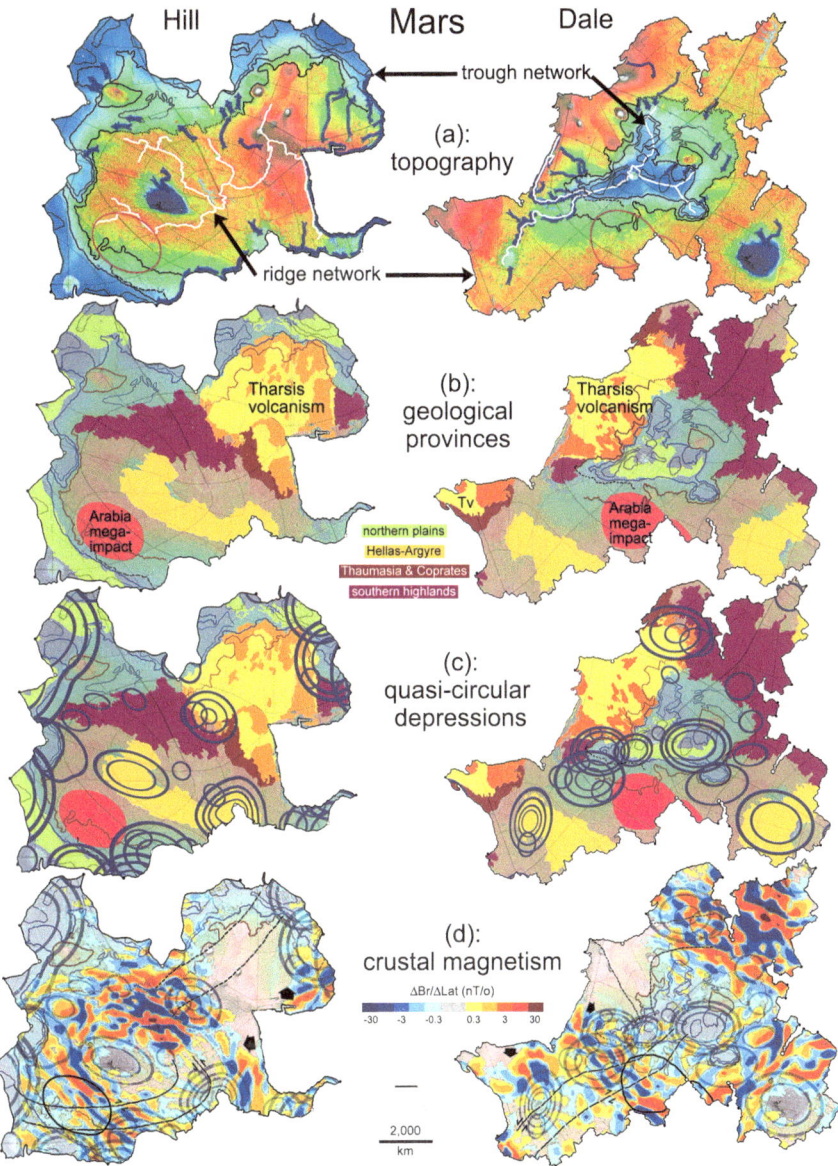

Fig. 5.5 Mars CSNB trough-bound (*left*) and ridge-bound (*right*) maps (Clark 2007) illustrating relationships between (**a**) topography (Smith et al. 2001) with boundaries indicative of shorelines (Parker et al. 2002), (**b**) geological provinces (Dohm et al. 2005), (**c**) large circular depressions (Frey 2012, personal communication), and (**d**) crustal magnetism (Connerney 2004) not clearly seen in conventional projections (Source data from which boundaries derived MOLA topography, courtesy of NASA and USGS)

topography due to differences in thermal isostasy in the pre-Tharsis era, subsequent erosion and other factors. Well-studied putative shorelines will deviate from this bench-line (shown dotted). Shoreline-highland overlap perhaps explains genesis of Medusa Fossae.

We indicate geological provinces (Dohm et al. 2005) (Fig. 5.5b) and display magnetic and gravitational anomalies (Fig. 5.5d), to illustrate potential underlying relationships.

We plot thirty-four primary mode large-scale visible and buried circular depressions with size/frequency distribution below 800 km (Frey, unpublished center and diameter data) (Fig. 5.5c). Thus we note that these features align with critical troughs, and almost all critical troughs run through them. Arcadia and Sirenum/Cimmeria basins stand out as the only dales not known to contain these features.

Crustal magnetism data (Connerney 2004) (Fig. 5.5d) were processed in the same manner as MOLA data. The close relationship between magnetic anomalies and the large circular depressions, described by Connerney as "intriguing" in global and geometric distribution, is clearly illustrated.

5.5 Future Applications

We anticipate further work on existing as well as new 'regular body' targets, as described below.

Earth: We are interested in further work with meteorological animations of the Earth as described in Sect. 4.3.

Mars: The present CSNB topographic maps of Mars could be pruned or extended. A more segmented (e.g., Fig. 1.8) would further distinguish features, similarities and differences among terranes and improve effective resolution. The ridge tree we show through Hellas could be grown to Hinges 9 and 10, and into Sirenum, Cimmeria, and Alba Patera (Hinge 14), clarifying their shapes and outflows. This is analogous to creating smaller-scale maps in orthodox cartography, but without the distortion of conventional global maps. Alternatively, more compact maps would unify features and emphasize global-scale differences. Pruning Points 4 and 5 off the tree forming the trough-bound map would unify Valles Marineris and Argyre Planitia. Pruning Points 14 and 15 off the tree forming the ridge-bound map would unify Solis and Tharsis. The ridge tree through Hellas could be grown to Points 9 and 10, and new branches extended into Sirenum, Cimmeria, and Alba Patera (Point 14) would clarify shapes and outflow points of these districts.

We could also generate maps from other ridge or valley trees.

1. A tree grown through medial axes of provinces remote to Tharsis and Elysium would generate a global map centered on volcanism, with antipodal geology as critical context. The complementary tree within the terrane of Tharsis magmatism, finely branched but regionally limited (similar to Fig. 2.7), would also

make an intriguing map. Global impact of this huge volcanic complex would be clear, because both lava flow and bulging strain would be inward (in terms of the map), and suppositions about super-plumes (Carr 2000) could be seamlessly entertained. A waterlining study (as described in Sect. 2.6) might tease into focus the antipodal effects of Tharsis-induced tilt on Meridiani sediments.

2. A tree grown within the southern highlands would emphasize the global impact of Aeolian erosion.
3. A tree grown through the medial axis of magnetic neutrality (conceptually a Maxwellian hill) would make a map focusing on magnetic crust, and display an uninterrupted magnetic geometry within a neutral frame of reference. Conversely, a tree grown through magnetic ridges (conceptually a Maxwellian dale) would focus neutral crust within a magnetically critical frame.

Mercury: We will begin work on Mercury as soon as the MESSENGER global-scale topography map, to complement the recently published global photomosaic, is completed. Geological features are 'subdued' and not always visually identifiable in the relatively high sun angle photographic coverage available for much of Mercury. Subdued features result from high temperature thermally driven global-scale deformation combined with large-scale episodes of volcanic resurfacing; thus, the availability of topography data to tease out low relief geomorphology is critical. Mercury is particularly interesting because, like Mars, it bears evidence of widespread externally driven as well as internally driven crustal modification processes. However, tidal despin and differentiation of the interior on Mercury, which has an Earth-like core, resulted in globally and not just regionally distributed volcano-tectonic features, including, in historical sequence, a global network of extensional faults and crustal expansion due to core expansion, followed by global volcanism, followed by a global network of compressional faults and crustal contraction due to core shrinking. Later episodes were likely to have reactivated earlier fault lines, as were the major impact events, which are typically associated with ray systems that are not strictly radial as they are on the Moon.

Outer Solar System Satellites: We intend to produce a set of CSNB maps for the inner Jovian satellites, to contrast their surface processes, ranging from internally driven, global-scale sulfur/silicate volcanism on Io causing rapid resurfacing, to a progressively smaller role for internally driven ice-based surface modification processes and greater role for external bombardment dominated surface modification as satellite distance from Jupiter increases.

Chapter 6
CSNB Mapping Applied to Irregular Bodies

6.1 Overview of Application

In 2001, geophysicist Dave McAdoo, viewing CSNB folded forms resembling Brazil nuts, observed that if a planet had been that shape, then the source map (Fig. 1.10c) would be an excellent 3D model for that planet. This led to CSNB mapping of asteroids (Clark 2002, 2003). When highly segmented CSNB maps of asteroids are folded, they produce reasonable, if not excellent, 3D representations; thus, CSNB mapping becomes a powerful 3D modeling, visualization and educational tool. We have used the CSNB approach to map and model asteroids to provide far more morphological insight than can be gained in the context of traditional flat (2D) map projections and regular plate (3D) models (Clark and Clark 2005, 2006a).

Why do cartographic methods produce such distorted maps of asteroids, a class of irregular objects? Cartography was devised during the centuries when map-making was applied to regular bodies: sphere-like planets. When a conventional map is presented for an asteroid, we should remember that the irregular body has been placed within a sphere, and its surface projected onto that sphere. Only then is it transformed to the plane, usually via a cylinder. The cylinder touches the asteroid only at the long axis ends; equatorial map-distance is not constant (Stooke 1998). CSNB maps, on the other hand, begin not with a sphere but with well-defined boundaries. A regular body's surface is relatively smooth, with variations in elevation small on the scale of the object. Segmenting that surface into terranes separated by ridges and troughs does indeed illustrate relationships resulting from surface shaping processes, but folding that model does not create an accurate 3D model of the object. On the other hand, asteroids are naturally irregular in shape, with large variations in elevation on the scale of the object; these objects are 'faceted' by bombardment of objects of more comparable size and too small for gravitational forces to transform into triaxial ellipsoids. Asteroid facet edges are boundaries in the dual sense of being impact event horizons, and being polyhedral on the scale of the object. Such boundaries may be imagined as a collection of hinges, resulting in a

P.E. Clark and C. Clark, *Constant-Scale Natural Boundary Mapping*
to Reveal Global and Cosmic Processes, SpringerBriefs in Astronomy,
DOI 10.1007/978-1-4614-7762-4_6, © The Author(s) 2013

foldable and representative 3D shape model. Conventional cartographic projections not only dismember but also obfuscate these shape-controlling boundaries. CSNB 3D models grow organically out of the most obvious facet 'edges' on an object, instead of being captive to a triangular 'plate' model that imposes regularly shaped facets on an irregular surface (Thomas et al. 2002). CSNB maps of asteroids, acting as ortho-normal frames, may be folded into photoreal models (Lakdawalla 2008).

We have collaborated to create CSNB maps for a series of asteroids with morphological modalities ranging, respectively, from clearly bimodal to more uniform, as illustrated, for example, by Castalia, Ida, Phobos, Deimos, respectively (Fig. 6.1). With the goal of establishing map-derived parameter indices for morphological complexity, we experimented with and developed a standardized CSNB approach for asteroid mapping, ultimately placing the blunt 'nose' at the center to allow systematic comparison of overall shape and degree of complexity on three axes (Clark and Clark 2005, 2009) (Sect. 6.3). We have also used the technique to clarify the relationship between deposition, erosion, rotational axes, and major impact events (Clark and Clark 2006a) (Sect. 6.2), to develop maps to be used for exploration of a typical, small near Earth object (Itokawa) (Clark and Clark 2010) (Sect. 6.4), and to study non-planetary irregular objects (Section 6.5).

6.2 Distribution of Features on Asteroid 433 Eros

We first applied the CSNB approach to Eros, creating antipodal conformal maps with major ridges (facet edges) as boundaries (Clark and Clark 2006a) (Fig. 6.2). Using NEAR topography as background, we plotted depositional ponds, and craters with eroded bright patches (Mantz et al. 2002; Cheng et al. 2002). We show two maps: one a compact map centered on the largest crater, Psyche, illustrating ridge structure relative to Psyche; the other highly segmented, centered on the sharp nose. Understanding morphological patterns requires minimizing distortion in important features, and CSNB maps are conformal for antipodal areas and preserve proportions in map interiors. Segmented projections preserve size and shape as well. A cylindrical map of Eros (Stooke 2012), Fig. 6.3, although familiar and thus instantly orienting, maintains neither proportion nor size, producing great distortion not only at higher latitudes (Krantz 1999), but also along the equator midway between the noses. Figure 6.3's equal area maps (Berthoud 2005) are similar to the compact CSNB map, but are two separate maps rather than one global map. A disadvantage in the segmented projection is the vigilance required to keep track of features (or graticles) for orientation.

The cylindrical projection (Fig. 6.2c) gives the impression that most of the significant regolith depositional and erosional features are found in the equatorial region (Mantz et al. 2002; Cheng et al. 2002).

CSNB maps allow relationships between noses, saddles, and poles to be observed without areal distortion. Topographic maxima and minima, representing bombardment history, align with map boundaries and thus are emphasized on CSNB

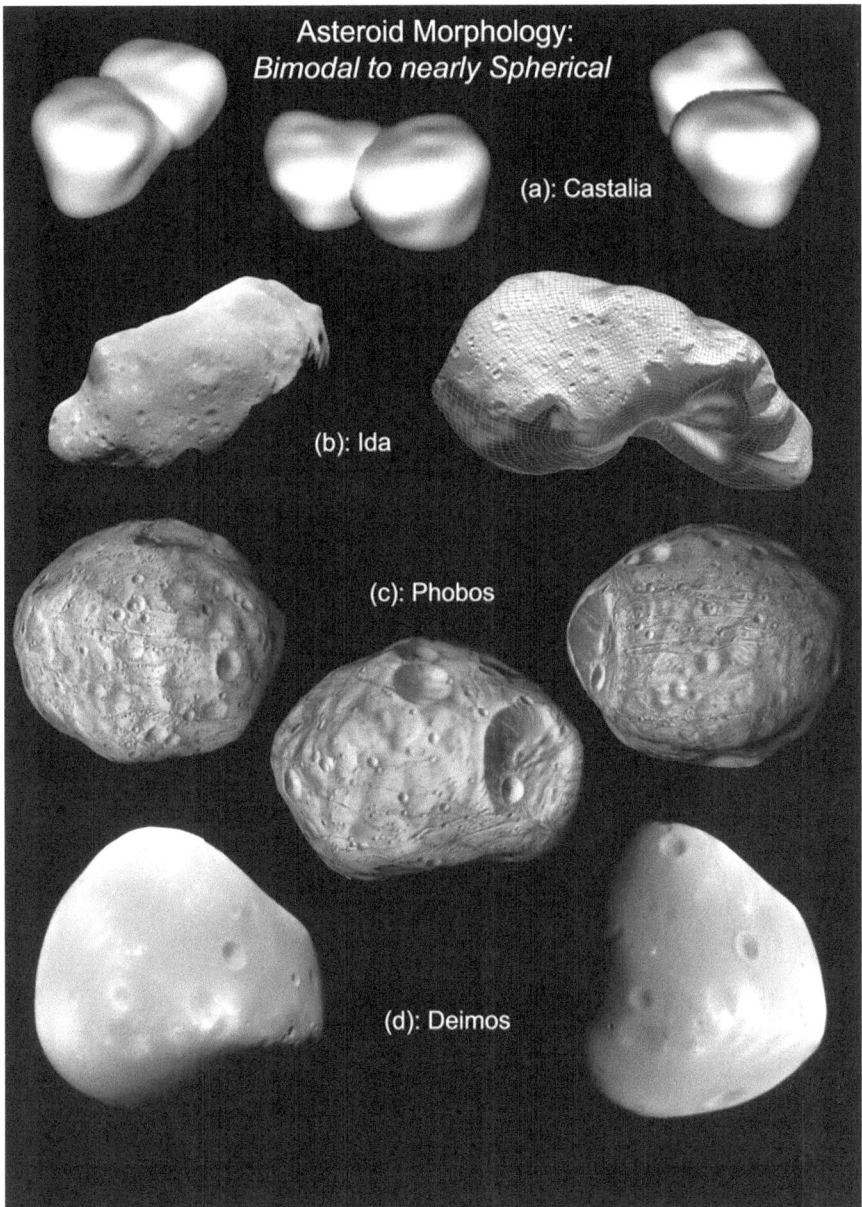

Fig. 6.1 Range of asteroid morphologies, as discussed in the text. (**a**) and (**b**) *left* Courtesy of NASA/JPL. (**b**) *right* and (**c**) Courtesy of Tayfun Öner. (**d**) Courtesy of NASA/JPL/University of Arizona

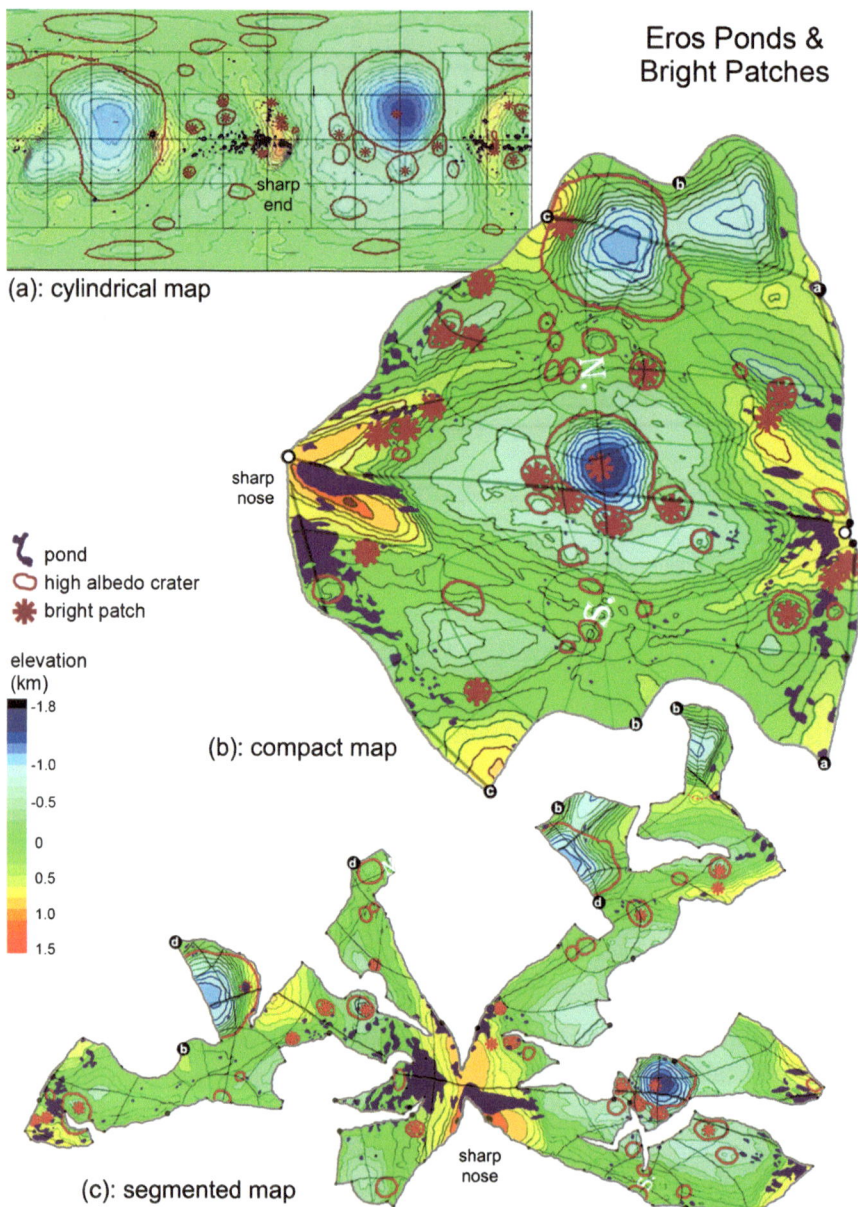

**Eros Ponds &
Bright Patches**

(a): cylindrical map

sharp
end

ʒ pond
◯ high albedo crater
✳ bright patch

elevation
(km)

-1.8
-1.0
-0.5
0
0.5
1.0
1.5

sharp
nose

(b): compact map

sharp
nose

(c): segmented map

Fig. 6.2 Compact (**b**) and segmented (**c**) versions of CSNB maps of Eros topography with dust ponds and bright patches superimposed (Clark and Clark 2006a). Simple cylindrical projection (**a**) included for comparison, as discussed in Sect. 6.2 (Source topography Cheng et al. 2002. Source ponds and patches Mantz et al. 2002)

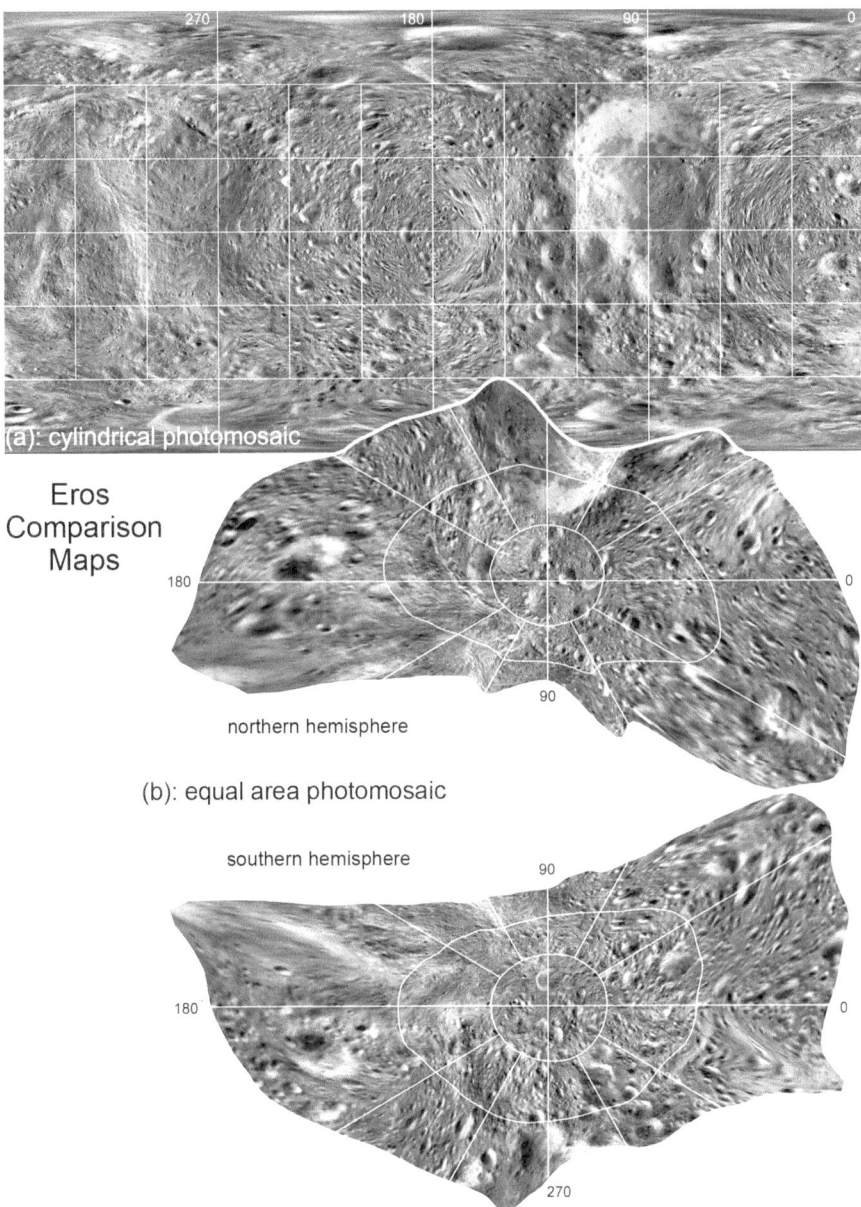

Fig. 6.3 Eros simple cylindrical and equal area photomosaics for comparison. (**a**) Courtesy of Phil Stooke. (**b**) Maps courtesy of Berthoud. Source Data Berthoud (2005)

segmented maps. The CSNB segmented map (Fig. 6.2c) clearly indicates that both features are found at considerable distances from the equator, even approaching the poles. Ponds are found near or on boundaries, particularly fanned out on the 'lee' side of noses, near local topographic maxima that apparently act as 'dust collectors'. Bright patch craters are found at all elevations surrounding the ponds, perhaps providing a source of the dust.

6.3   Comparison of Eros, Phobos, Deimos, and Ida

We used CSNB mapping of Deimos, Phobos, Eros, and Ida to perform a systematic study of asteroid shape and surface morphology. As in the Eros study, the ridges and troughs were seen as event horizons acting as encoders of asteroid history, appropriately identified as critical boundaries and thus prominently featured as map edges. For consistency and orientation, we centered all maps on the blunt nose, i.e., we pruned the tree from the blunt nose, because most asteroids elongate along one equatorial axis and the blunt nose is a recognizable feature, but, as we learned in the previous study, less complex than the sharp nose. The external boundaries became ridges, which typically run parallel to the equator, and troughs, which typically separate promontories and wrap the 'waist'. Three maps, two ridge-bound (one compact, the other segmented) and one trough-bound (compact), were created for each object (Figs. 6.4 and 6.5). Segmented maps (middle row) separate the surface into facets, preserve size resolution, and fold to excellent 3D facsimiles. Compact maps (top and bottom rows) preserve orientation and are suitable for customary map uses, with the added property of physical meaning at their edges. Compact maps also fold, and the resulting volume may be roughly congruent to the originating object, as we saw in Fig. 1.5, but, as we saw in Fig. 2.7, the folded form may be unsympathetic, and meaningless beyond formal considerations, such as a check on drafting accuracy.

Understanding morphological patterns requires the ability to minimize distortion in important features and to see relationships for the entire surface simultaneously on one map. Thus, although we use them for comparison and orientation (Figs. 6.6 and 6.7), we eschew standard projections and 3D views. CSNB maps combine the best of those traditional geometric demonstrations, emphasizing irregular faceted shape in one, unitary, minimally distorted figure. Our CSNB maps show crater distribution as well as radial distribution of albedo and other morphologies relative to the 'pruned feature', which in our study was the blunt nose, but of course could be any spot of overarching interest.

Morphological parameters manifested in CSNB map shape include E/W and N/S distribution of segments, roughness of boundaries associated with each segment, and aspect ratios for each segment. Based on comparison of these parameters (Table 6.1), Phobos and Deimos are more spherical and regular than Eros or Ida, implying a more disruptive history for the latter two asteroids. Phobos is composed of considerably elongated segments, due to the network of radial grooves extending

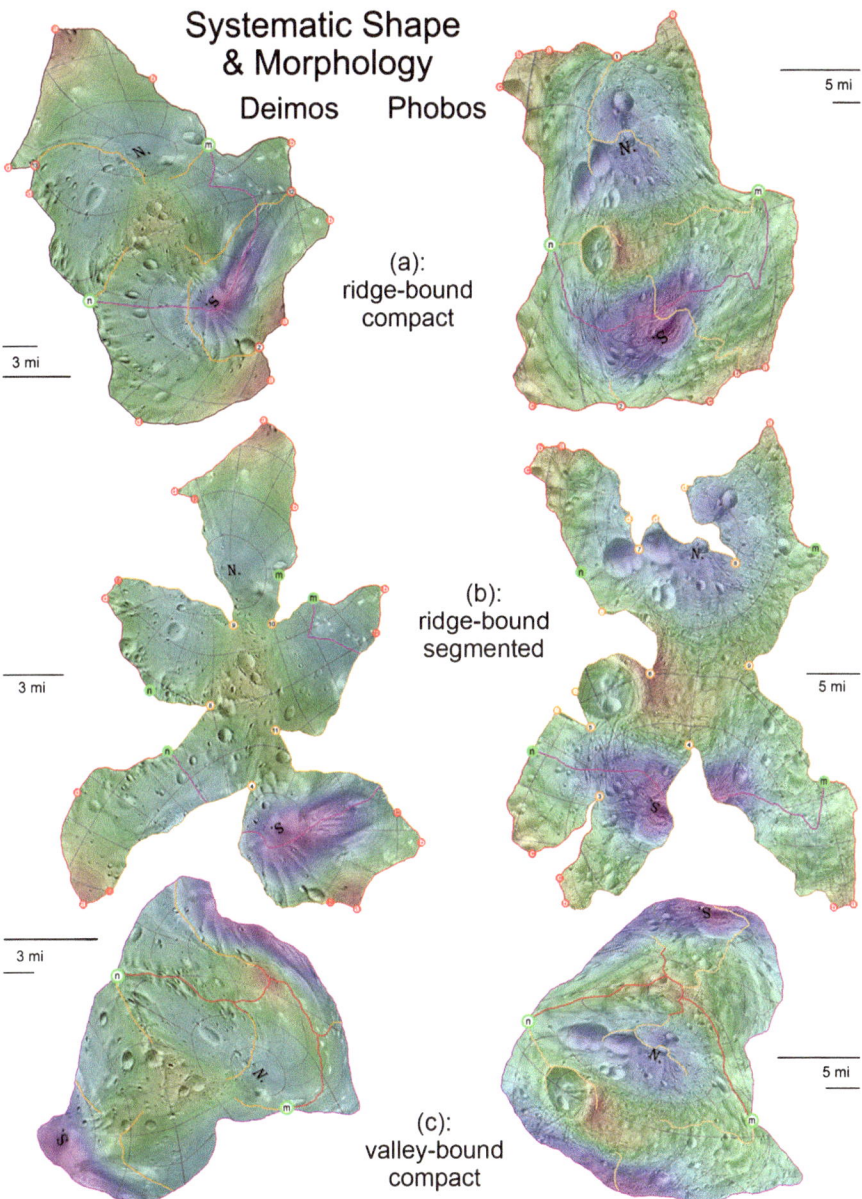

Fig. 6.4 Variety of CSNB maps for Deimos and Phobos to illustrate shape and morphology systematics (Clark and Clark 2009) as discussed in Sect. 6.3. Topography data courtesy of Tayfun Öner

Fig. 6.5 Variety of CSNB maps for Eros and Ida to illustrate shape and morphology systematics (Clark and Clark 2009) as discussed in Sect. 6.3. Topography data courtesy of Tayfun Öner

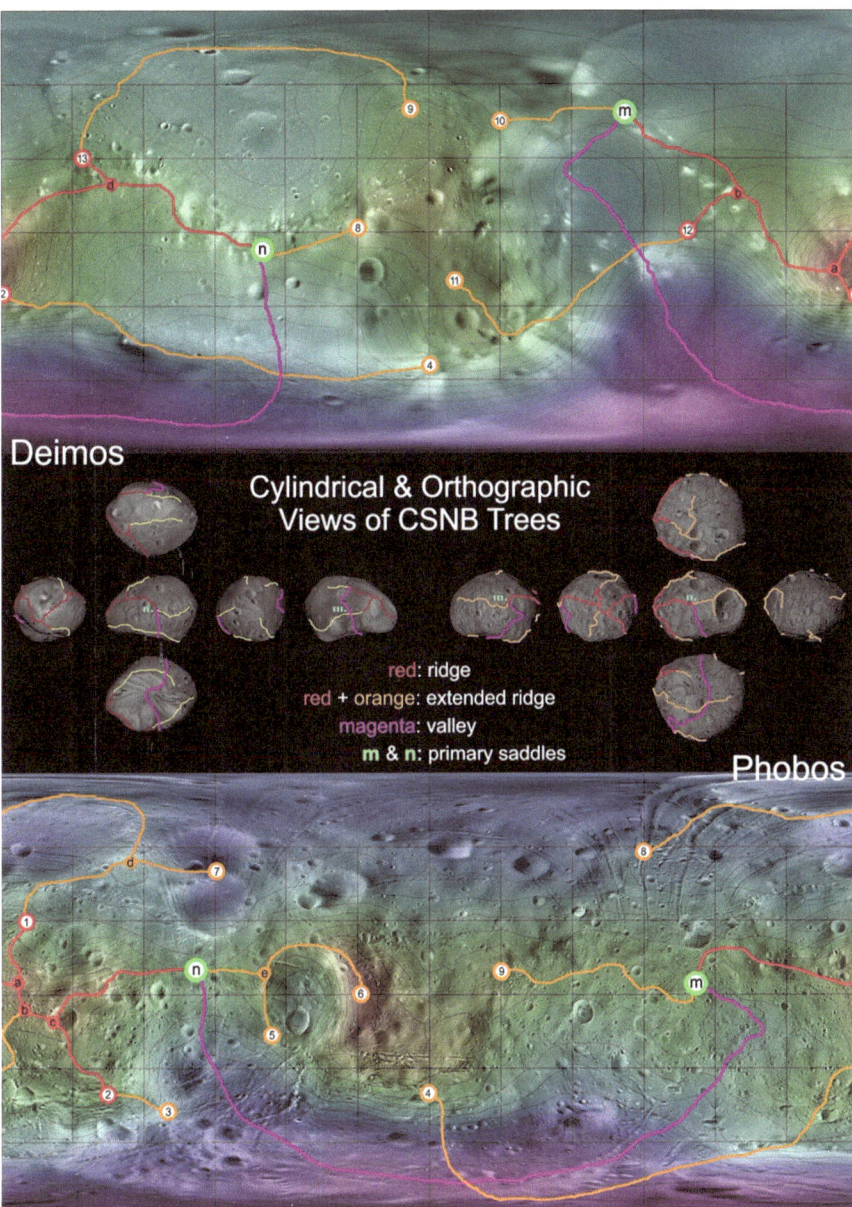

Fig. 6.6 Phobos and Deimos with ridge and valley lines superimposed on 3D models and simple cylindrical maps for comparison (Clark and Clark 2009) (Source: Deimos photomosaic courtesy of Phil Stooke, Chris Jongkind, and Megan Arntz. Source Phobos photomosaic courtesy of Phil Stooke. Orthographic views and topography data adapted from Tayfun Öner)

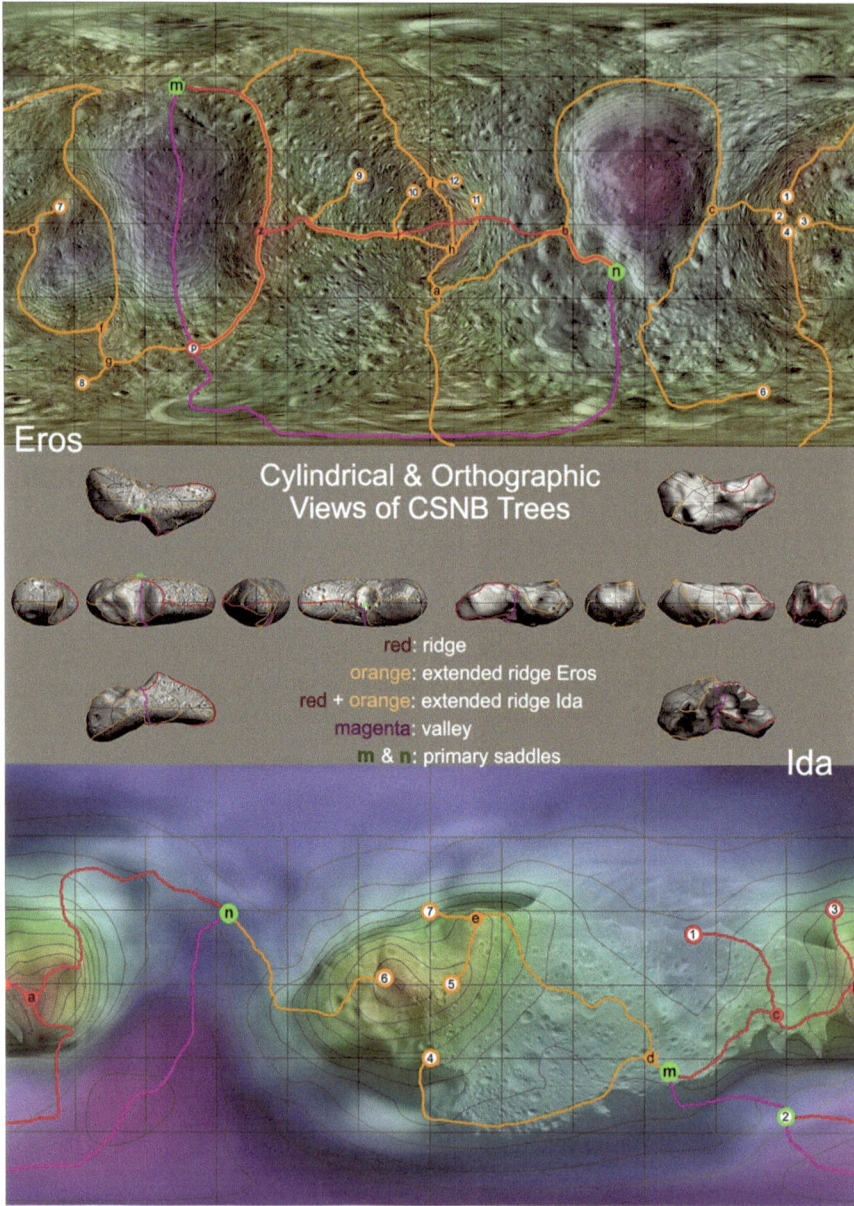

Fig. 6.7 Eros and Ida with ridge and valley lines superimposed on 3D models and simple cylindrical maps for comparison (Clark and Clark 2009). Source Eros photomosaic courtesy of Phil Stooke (Source Ida photomosaic courtesy of Phil Stooke and Maxim Nyrtsov. Orthographic views and topography data adapted from Tayfun Öner)

Table 6.1   CSNB derived asteroid parameters

Asteroid (size, km)	E/W symmetry N/S symmetry	Boundary roughness segmentation	Aspect ratio for segments (length/width) average, range
Deimos	E/W 2.5:2.5	11–12/facet	1.0 (0.5–2.0)
$15 \times 12 \times 10$	N/S 3:2	5 facets	
Phobos	E/W 2:3	15–27/facet	2.4 (1.0–3.0)
$27 \times 22 \times 19$	N/S 2:3	5 facets	
Eros	E/W 5:4	10–31/facet	1.2 (1.0–3.0)
$33 \times 13 \times 13$	N/S 7:2	9 facets	
Ida	E/W 3:4	7–24/facet	1.7 (1.0–3.0)
$54 \times 24 \times 15$	N/S 4:3	7 facets	

from the Stickney Crater on the nose. Deimos and Eros show considerably greater asymmetry in the N/S directions. Eros and Ida both exhibit considerably more roughness on segment boundaries, indicating that Martian moons Deimos and Phobos, with their less-disruptive facet boundaries, are somewhat shielded from bombardment by the planet Mars. Asymmetry in Deimos shape is caused by the sharp, southward hooking nose in the equatorial region and the cavity, presumably formed by an impact event, near the south pole.

6.4   Exploring Asteroid 25143 Itokawa

Having utilized Constant-Scale Natural Boundary mapping to characterize the shape and surface morphology of a range of asteroids as a key to understanding and comparing their underlying structure and history, we applied this technique to exploration of a typical small near Earth object, Itokawa.

Itokawa is one of thousands of the potentially hazardous asteroids with Earth-crossing orbits (Minor Planet Center 2012) that are considered prime candidates for exploration. Observations made by the Hayabusa mission indicate that Itokawa is an excellent analogue for such asteroids (Landau et al. 2011) due to its small size ($0.7 \times 0.7 \times 0.3$ km), low density (1.95 g/cm^3), extreme ruggedness and rubbly surface, supporting its interpretation as a rubble pile (Hirata et al. 2008). Its peculiar 'ring around the collar' inflection has created additional interest.

How prominent, distinctive, and globally interconnected will the ridge and trough boundaries be on a rubble pile? How can we use that information in planning exploration routes over or on the surface?

Itokawa is an asymmetrically bimodal asteroid covered predominantly by rough, boulder-bearing terrain and to a lesser extent by smooth, in-filling dusty terrain (Hirata et al. 2008; Miyamoto et al. 2007). The latter includes quasi-circular dust-filled depressions, or ponds, without rims and troughs, as observed on Eros. Some ponds may be craters penetrating the shallow regolith. The dominant morphological feature is a band or trough forming a 'neck'. This feature and less-obvious shelves or shoulders at both ends, as well as small ridges running from nose to nose, have most likely resulted from loss of material during dynamic interactions. Ridges and troughs are often locally prominent, but discontinuous, and not clearly associated with other impact features. Downslope migration of debris to low-lying areas exposes material and occurs on a more rapid scale than space weathering (Hirata et al. 2008; Miyamoto et al. 2007).

In order to create operational scenarios for exploration of this analogue, we have studied photomosaic, albedo, color imagery, 3D models, and a simple cylindrical shaded relief map (Figs. 6.8). The overhanging neck complicates orthodox cartography: each ray identifies more than one surface point. This is not a problem in CSNB because rays are not used in the flattening transformation.

From these we derived valley-bound CSNB maps centered near the Hayabusa sampling site known as the Target Marker (as seen in Fig. 6.9). Topographic boundaries (depressions, ridges, craters, and valleys), potential sampling sites (rocky ridges and boulder fields, depression regolith pools, crater rims, prominent structural inflections) and traverses connecting these sites are indicated. Hundred- meter-diameter circles drawn at various latitudes indicate the systematic and non-systematic distortion in asteroid distances on standard geographic projection maps and the difficulty in using them as field maps.

We use the CSNB method to provide relevant information on morphological and geographic relationships including E/W and N/S distribution of segments, roughness of boundaries associated with each segment, and aspect ratios for each segment. Among asteroids we have studied to date, Itokawa is the most irregular, and the most clearly bimodal, implying the most disruption. Itokawa also exhibits the most asymmetry in the N/S and E/W directions.

We derived a route map that would follow ridgelines (Figs. 6.8 and 6.9), allowing the broadest perspectives and views of sites in context, as well as visibility and safety of team members. Approximately fifty potential sampling sites, associated with prominent structural features including noses and poles, boulder swarms or regolith pools, occurring along natural topographic breaks or not far from them, are identified. Traverses, hundreds of meters long, generally follow these breaks.

Each site requires up to an hour of tethered sampling and other activities. Assuming travel between sampling sites is analogous to rock climbing (tether/rappel-tether/release), we estimate travel between sites would occur at about three meters per minute. This translates into about four to five sites a day, or the capability to visit all fifty sites in about two weeks. The availability of a jetpack scooter could speed up travel between sampling sites, allowing more sites to be visited or more time to be spent at sampling sites.

Thus, the highly segmented, nearly orthographic CSNB maps are particularly useful for route planning, typically along ridges. The CSNB map(s) show districts by horizons as they would be seen by an explorer, and could adapt to changing conditions and mission objectives, e.g., if the nose were a target (we avoided it), additional map-interruptions into this district would bring its 100 m circle up to scale with the others in Fig. 6.9, and lengths of potential nose-traverses could, like the other routes, be read directly.

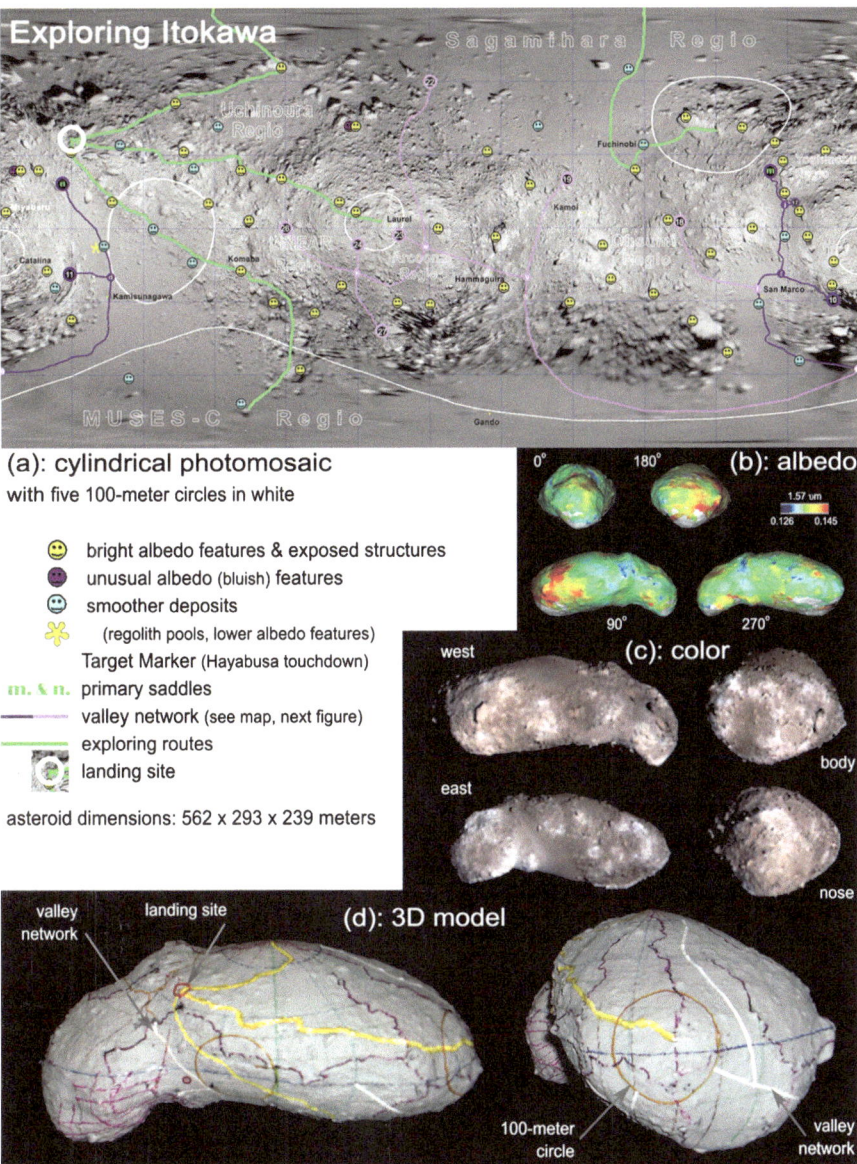

Fig. 6.8 Itokawa features and exploration routes in conventional projections and views (Clark and Clark 2010). ((**a**) Photomosaic courtesy of Phil Stooke, Images (**b**) and (**c**) courtesy of JAXA/ Hayabusa Project Team, (**d**) Plastic model courtesy JAXA/Hayabusa Project Team and photos courtesy of Sara Adkins Studio)

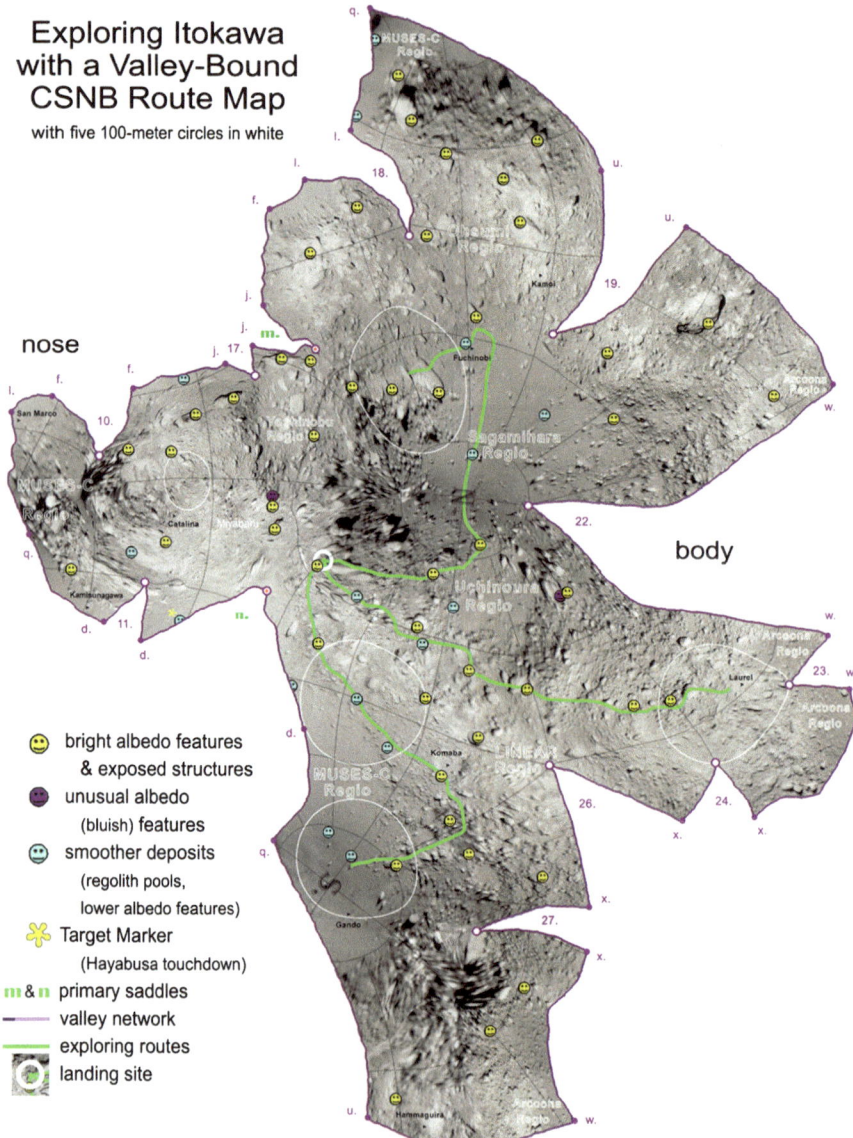

Fig. 6.9 Itokawa features and exploration routes on a CSNB map (Clark and Clark 2010) (Source photomosaic courtesy of Phil Stooke)

6.5 Other Irregular Objects

CSNB precisely maps the surface geometry of objects with irregularity commensurate with their scale, a virtue when studying natural objects. Here we show aspects of living organisms (hominid crania) and molecules. Although this book is entitled "in the solar system and beyond," these objects and how we study them have inductive relevance to planetary and astronomical objects and events. Certainly molecules and perhaps living things are distributed throughout the universe.

Figure 6.10 illustrates comparative cranial mapping, not only useful for evolutionary anthropology but also a prototype for our future comparative Jovian moon study (Sect. 5.4). Because crania are topologically complex objects (i.e., high genus or many-holed), the fact that CSNB can map them with foldable accuracy bodes well for digital CSNB map simulations of analogous maps for galactic events (e.g., colliding galaxies). Figure 6.11a makes the case for ranking CSNB as a new class of geometrical demonstration, on par with Renaissance-era perspectival views and Euclidean cross-sections and plans. Figure 6.11b maps a water molecule. Formation of analogies between processes involving fundamental particles and cosmic processes have a long history, and the quantum world of the atom and quantum fluctuations of the cosmic microwave background radiation are based on the same fundamental physics. (In Chap. 7 we map the cosmic microwave background.)

Crania: Fig. 6.10 shows three hominid specimens, maps, and centerlines. Maps and centerlines are compared by overlay, as architects do with floor plans. Specimens vary in size but maps (and photos) are scaled to equal longitudinal length; the foramen magnums (spinal column hole) are fulcrums to align and orient, as we used asteroids' blunt noses in Sect. 6.3. The Neanderthal is a solid cast; other maps are pruned to match. Overlays (**d**) and (**e**) reveal exact divergences in surface. Centerlines (pruned medial axes) reduce each cranium to a one-dimensional curve expressing both bilateral symmetry and asymmetry. The subtleties of disproportion about the midline of human facial appearance are the raw material of artists, distinguish us from robots, and make us attractive (Zaidal and Hessamian 2010); these subtleties are rooted in cranial disproportions, which we condense to a line with a characteristic S curve. Two specimens lean to the left and one, the Neanderthal, leans to the right. The inversion may be a mirror-artifact of casting, but if not, our results (though sample size is ridiculously tiny) suggest CSNB maps may identify handedness preference.

Extrapolating from this small study, we speculate that overlay analysis as a general technique may give probabilities of familial lineage in a specific cranial population, or, more broadly, assess whether bilateral deviations are random or systematic. This may be a fruitful niche of inquiry in evolutionary and forensic anthropology, and we look forward to similar experiments with Jupiter's family of tidally locked moons.

CSNB crania maps have no precedent since experimental anatomy began in the fifteenth century with Leonardo da Vinci. Software and other technologies have made visualization easy yet, while intervening centuries brought non-Euclidean

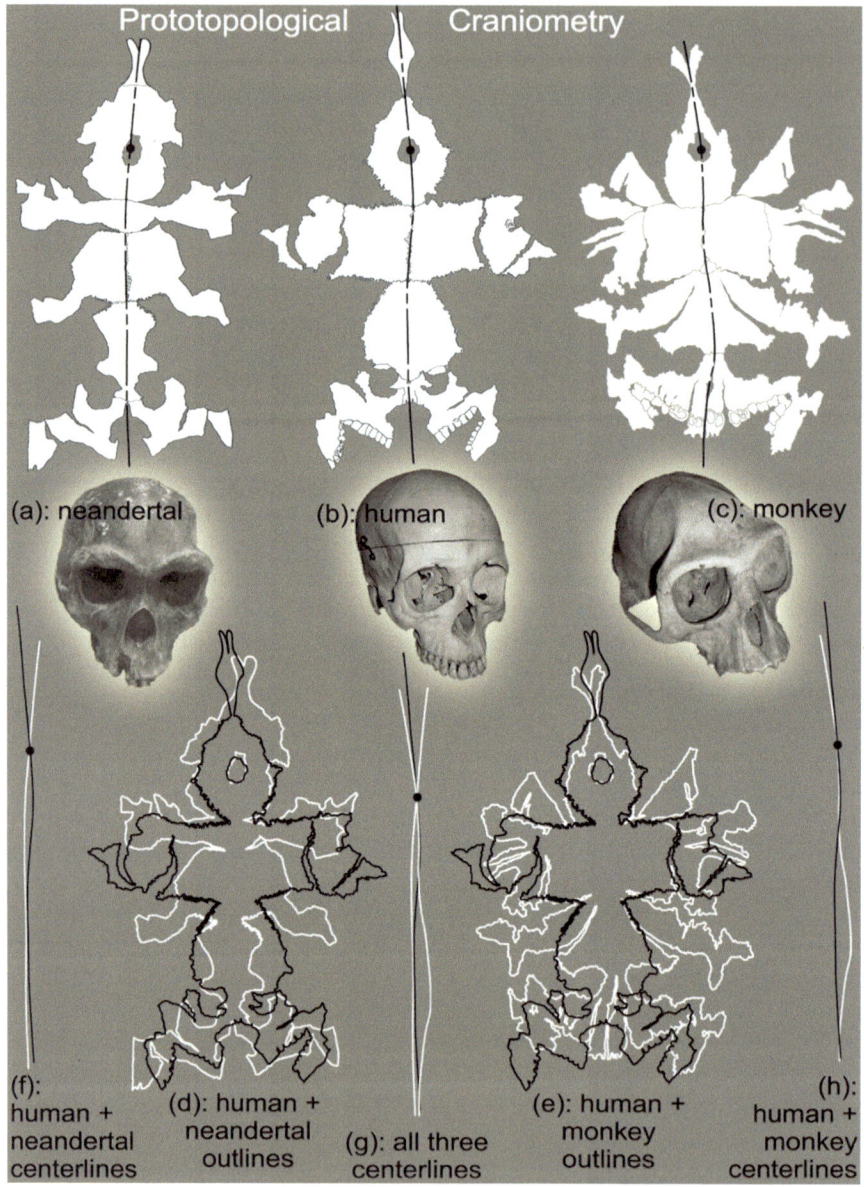

Fig. 6.10 Comparative cranial mapping using CSNB and medial axis techniques. Sources (**a**) Broken Hill #471 cast specimen courtesy Glen Doran specimen (**b**) and (**c**) courtesy of Susan Clark Williams collection. Crania photos courtesy of Sara Adkins Studio

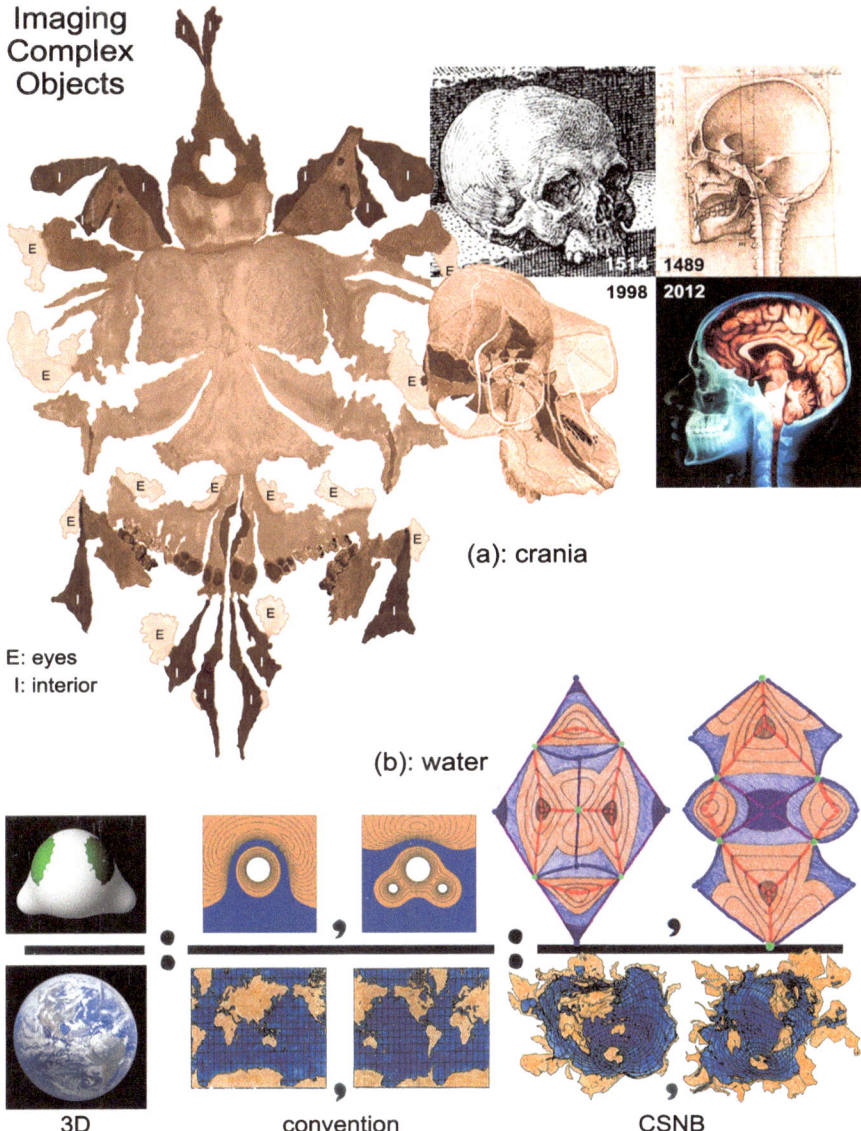

Fig. 6.11 Comparison of CSNB versus conventional viewing of complex objects such as crania, water molecules and Earth, as discussed in Sect. 6.5. Sources (**a**) CSNB monkey cranium Clark (2003), skulls from Dürer (detail from *St. Jerome in his Study*) (1514) and da Vinci (detail from *Two Views of the Skull*) (1489); skull cross-section courtesy of Steven J. Goldstein, MD, and the MedPix® online database; (**b**) 3D Earth Courtesy of NASA, CSNB Earth Clark (2002); 3D and conventional water, Young and Harrison, CSNB water: created by Clark and Clark in 2013

geometries and topology into existence, our output paradigms remain cross-sections, plans, and perspective verisimilitude (Keele 1983) (Fig. 6.11a). The CSNB monkey-cranium map shows considerable internal surface, and if digital measurement capture had been available (measuring with wires takes you only so far) could have propagated to show the full interior of the brain vault. We argue that CSNB is only superficially related to conventional projections; instead it is a non-Euclidean analog of architectural plans.

Water Molecule: The 3D configuration of molecules is a most significant characteristic. Electron density isosurfaces control access and interactions of the molecule with external agents. Surfaces of molecules are now being 'mapped' in exquisite detail with cross-sections and 3D views, which yield soft, distorted, or incomplete representations of the molecule's actual, three-dimensional shape (Fig. 6.11b). Crucial spatial subtleties such as a free water molecule's tetrahedral lone-pair electron bulges and consequent, molecular-plane electron shallows (in green at *left*), are not easy to discern, and a complete awareness of the molecule's surface must rely on the mind for synthesis.

The effective shape of a molecule, the availability of a surface for chemical interaction, could be expressed as electrical potential at a given distance, controlled by effective electron density, analogously to potential gravitation energy correlated with elevation in the form of peaks, pits, and saddles on the surface of a planetary object, as described by Maxwell (1870). In the molecular case, contours are isobars of classical form carried as second order functions on the isosurface. The variable, contour-creating reference volume could be a sphere about the center of mass, a triaxial ellipsoid about the center of mass, or, for a three-atom water molecule, a tri-focal constant-total-radius volume, similar to that given by Maxwell (1846) for a planar figure of three foci: $mr^1 + nr^2 + pr^3 =$ constant. With these contours of the isosurface, we find ridge networks connecting density maxima through saddles, and trough networks connecting density minima through bars. These networks lead directly to CSNB maps (Fig. 6.11b), which may be thought of as plats of molecular real estate (Krantz 1999).

Thus, CSNB conformal maps of electron density would give recognizable structure and pattern to descriptive critical boundary.

Chapter 7
Mapping the Sky

7.1 Beyond Human Sight

Maps of the sky abound. These include the familiar 'night sky' maps of visible constellations, which we won't discuss here, as well as sky maps from major regions of the electromagnetic spectrum (The MultiWavelength Sky):

1. NRAO Radio Sky at 73.5 cm
2. NASA COBE and WMAP Microwave Background at 1.87 mm
3. NASA COBE Diffuse Infrared Background Experiment at 240 μm
4. NASA COBE Diffuse Infrared Background Experiment at 3.5 μm
5. NASA HEAO-1 soft (2–10 KeV) X-rays
6. NASA Compton Gamma-ray Observatory map at 100 KeV.

Here we focus on the cosmic microwave background, with some mention of the infrared background because of its significance to the origin of the universe.

7.2 Cosmic Microwave Background

The cosmic microwave sky background is thought to be radiation left over from the earliest stage of development in the universe (the Big Bang) when the universe was hotter, more uniform, simpler in structure, and bathed in elementary particles. Cosmic microwave background (CMB) photons interacted with matter until cool enough conditions persisted so that ionized plasma components could combine to form a range of elementary particles allowing photons to escape and scatter in all directions. As the universe evolved, it continued to cool and develop a more complex structure. The escaped CMB photons, although at earlier times the dominant part of the mass-energy budget, now account for only a small fraction of that budget, but have a recognizable thermal black body spectrum temperature, now about

P.E. Clark and C. Clark, *Constant-Scale Natural Boundary Mapping*
to Reveal Global and Cosmic Processes, SpringerBriefs in Astronomy,
DOI 10.1007/978-1-4614-7762-4_7, © The Author(s) 2013

2.725 K (Mather et al. 1999), the equivalent of about 1.87 mm wavelength in the microwave region.

As evidenced by all-sky maps from first the COBE mission and then the more sensitive and higher resolution WMAP mission, when this spectrum is measured in all directions, it is not uniform (Fig. 7.1), indicating initial heterogeneity in the distribution of matter. The interaction between the pressure driving CMB photons outward and the inward force of gravity created acoustic oscillations now reflected as fluctuations on the maps and observed in their spherical harmonic power spectra (Hinshaw et al. 2007). Small polarization effects, resulting from elastic scattering from non-isotropically distributed electrons (Kovac et al. 2002), were also anticipated and observed. An interesting feature of these maps, also observed on COBE diffuse IR background maps to some extent (Fig. 7.1a) is the similar pattern in the apparently asymmetric distribution of pronounced low temperature and high temperature regions, which we will discuss below.

COBE confirmed the existence of the thermal background radiation and the predicted oscillations from uniformity. The WMAP map (Bennett et al. 2011), which measured differences in temperature, played a key role in establishing the current standard model of cosmology (Lambda-CDM Model). WMAP observations confirm the age (13.75 BY) of the universe, the cosmic inflation paradigm and the Hubble expansion constant, a Euclidean flat geometry with predicted ratio of energy density in curvature to critical density, and predicted proportions of dark energy (dominant) and dark matter. However, the CMB maps also display anomalies, relative to the standard model predictions, including larger-scale anisotropies such as pronounced cold spots, and non-random alignments of spherical harmonics (confirmed for lower orders, including second order quadrupole, third order octupole) about the ecliptic plane (Diego et al. 2010; Kim et al. 2012; Copi et al. 2006, 2010). According to Kim and coworkers (2012), the CMB map shows an overall odd-parity (asymmetry) preference in disagreement with even-parity (symmetry) assumed by the standard cosmological model. Copi and coworkers (2010) noted quadrupole and octupole planarity, mutual alignment, north/south asymmetry, and alignment perpendicular to the ecliptic and the dipole, and not the galactic or supergalactic planes, a result that they found "puzzling and intriguing". The ecliptic neatly separates adjacent cold spot (north) and hot spot (south). They also noted the greater amplitude for octupole (smaller-scale) than quadrupole (larger-scale) measurements and lack of two-point correlation (clustering) by points separated by more then 60°. It should be noted that at least some of these anomalies with their unanticipated distribution could result from incomplete removal of foreground 'contamination'. Diego and coworkers (2010) detected an additional signal with a quasi-blackbody spectrum with spatial distribution similar to zodiacal light correlated with the ecliptic plane. Other explanations, such as greater abundance of dust grains in the ecliptic plane, would generate very different results from those observed (Copi et al. 2010). Measurements of absolute temperatures rather than temperature differences, and polarization maps from the Planck experiment will be useful in resolving these controversies.

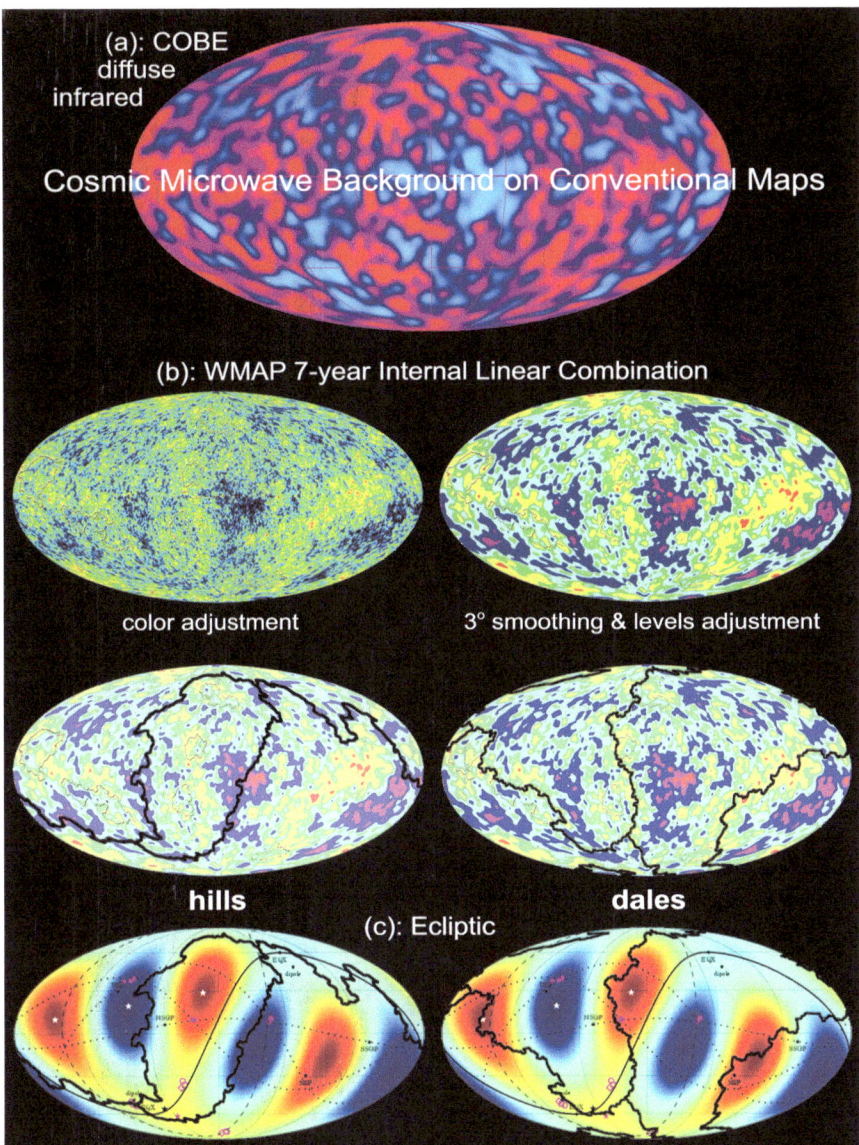

Fig. 7.1 Cosmic microwave background maps with hill and dale networks as described in the text. (**a**) Source data courtesy of NASA COBE. (**b**) Source data courtesy of NASA LAMBDA/ WMAP; (**c**) Adapted from Fig. 2, Copi et al. (2006) © The Authors. Journal compilation © 2006 RAS

7.3   CSNB Maps of CMB Anisotropy

Sky maps are usually in the pseudocylindrical Mollweide projection, including those in Fig. 7.1. This is an equal area projection that sacrifices accuracy of angle and shape for accurate areal proportions. The Mollweide sky maps are projected on our (Milky Way) galactic latitude and longitude grid. Note, however, the plot of the ecliptic, which is NOT the Milky Way galactic plane. In order to characterize the nature of lower and higher temperature regions, and their relationship to each other, maintaining accuracy in shape will be crucial. Thus, using the WMAP data, we create a series of CSNB hill and CSNB dale maps. The Hill series, Fig. 7.2, is bound by temperature increase minima; the Dale series, Fig. 7.3, is bound by temperature increase maxima. Both series progress generally from more to less segmentation, and end with a 'compromise' map, similar to the finely branched map in Fig. 1.8, that restores local peripheral structure to the most compact map in each series. Note, as seen in the Mollweide maps (Fig. 7.1c) we have three segments of anomalous cold or hot regions arranged triaxially. Both series organize into large, medium, and small segments. Compare the plot of the ecliptic, with which anomalies appear to be aligned, and the supergalactic plane, which forms a closed loop on the compact CSNB maps.

While it is beyond the scope of this work to fully report our study, our crawl around the contours of the cosmos identified six (and only six) districts of an insular nature, upland basins, or 'coves', comparable in size to cold spots identified by Bennett et al. (2011). These six districts are distinguished by extreme 'height' of encircling ridges, noisy, low-lying floors, and incised, singular spouts. They are neither extremely cold nor extremely hot, i.e., they do not contain the CMB's hottest peaks or coldest pits, but their special aggregation of hot and cold topography, distinct from the rest of the CMB, commends them to the attention of cosmologists. These coves are lightly embossed on the CSNB maps. Five coves are found north, and one south, of the ecliptic, another non-random result that is not easily explained.

Cosmic Microwave Background:
Hills

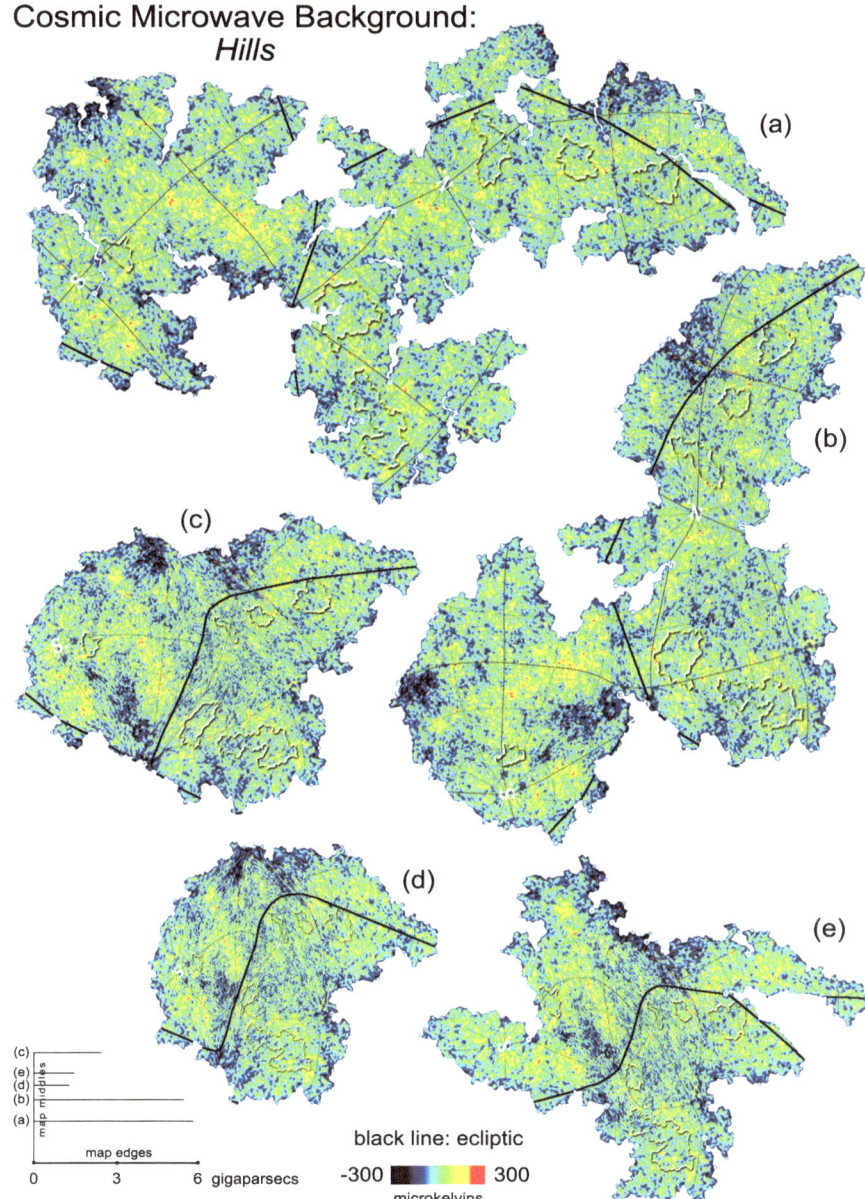

Fig. 7.2 CSNB maps of CMB hills with progress (**a–e**) segmentation sequence. All maps have the same edge scale (Source data courtesy of NASA LAMBDA/WMAP)

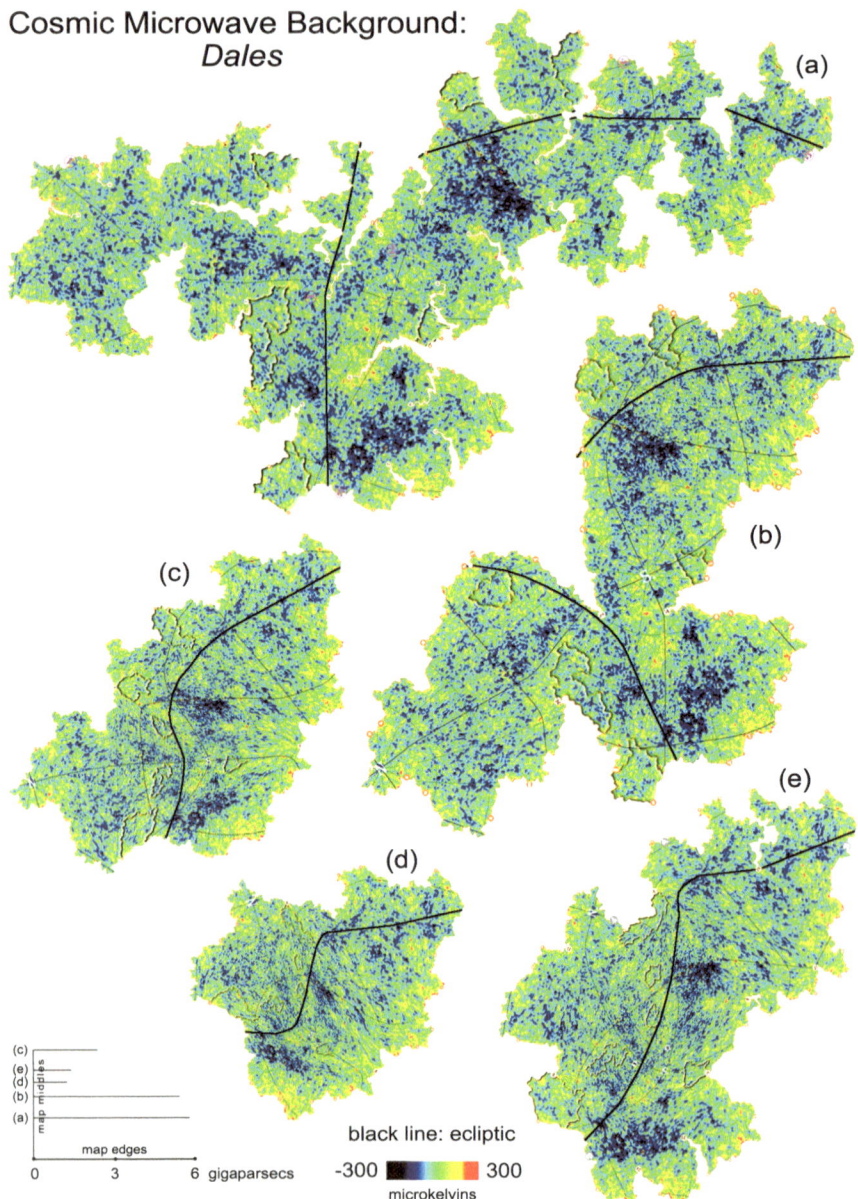

Fig. 7.3 CSNB maps of CMB dales with progressive (**a–e**) segmentation sequence. All maps have the same edge scale (Source data courtesy of NASA LAMBDA/WMAP)

Chapter 8
The Future of CSNB Mapping

8.1   Status, Goals, and Motivation

In the previous chapters, we have described Constant-Scale Natural Boundary mapping, a revolutionary paradigm for 2D and 3D visualization. The technique produces maps markedly different from, and yet complementary to, those produced by conventional 2D cartographic and 3D modeling techniques. Whereas conventional maps are built on predetermined grids or formulae which often distort prominent features that result from underlying processes, CSNB maps begin with well-defined boundaries, often found at the edges of conventional projections, that result from natural processes. By directly illustrating spatial and dynamical relationships, CSNB maps provide greater insight into formation processes.

Our goal is to automate our currently manual technique for identifying regional-scale extended maxima or minima trees in well-defined, global-scale parameter space (e.g., land or sea elevation, field anomalies, pressure, temperature, chlorophyll production, density). In this way, we can produce a range of 2D maps easily separated into segments (e.g., plates, vegetation zones, prevailing wind systems) defined by the range of parameters (hills and dales) for selected processes. We can project that data as 3D models. We currently seek support for this effort, which we see as occurring in several stages. In the first stage, the automated process would be used to make CSNB maps of regular bodies (Earth, moons, planets); in the second stage, the automated process would be modified for application to less regular objects (such as asteroids, nebulae, galaxies) where impact or other event-defined edges act as boundaries. Maps of such objects could be built on a 2D control net (as on regular bodies) but with a third dimension analogous to digital elevation maps for Earth and other triaxial ellipsoids: a spherical coordinate system with distance to center of mass to identify ridges and troughs as large angular variations for a highly irregular surface.

By making 'a new way of seeing the world' more accessible and portable, automated CSNB mapping would complement existing cartographic approaches

P.E. Clark and C. Clark, *Constant-Scale Natural Boundary Mapping*
to Reveal Global and Cosmic Processes, SpringerBriefs in Astronomy,
DOI 10.1007/978-1-4614-7762-4_8, © The Author(s) 2013

established and enshrined for over 400 years as well as the more recent, computationally intensive 3D plate modeling. CNSB mapping will significantly enhance the portfolio of mapping, image processing, and modeling techniques available to scientists and engineers. Automated CSNB will allow more rapid and comprehensive interpretation of global systems, displayed with greater continuity of the natural phenomena, and create the basis for faster, less labor intensive, 'state of the art' autonomous mapping and modeling of bodies and systems. Interest in this technique has already been expressed by our colleagues, including educators, in terrestrial oceanography, meteorology, and geophysics, as well as in Martian and asteroid studies, as described in earlier chapters. CSNB approach will help to enable future distributed survey missions, with onboard intelligence for 'target of opportunity' exploration.

8.2   Methodology and Plan for 2D Mapping

In phase one, the automation process would be used to create maps of regular bodies for which manual versions already exist, including Mars ridge network, and Earth tectonic ridge maps (Clark and Clark 2005; Clark et al. 2006). In phase two, we would modify the algorithm for irregular objects by incorporating spherical coordinates and associated distance to the center of mass, and automating the identification of object-scale features and angular variations as boundaries. Steps in this process include:

1. Developing mapping algorithm;
2. Obtaining the highest resolution digital elevation maps;
3. Assessing applicability (e.g., Lagerstrom et al. 2008; website tool), selecting, and testing branching ridge (watershed) or trough (watercourse) network recognition filters and applying them to find boundaries for each test case;
4. Performing surface interruption along constant-scale boundaries, using hinges and elbows to maintain internal proportions, and create segmented and compact versions for each map series;
5. Applying this approach to test cases for Earth climatology, meteorology, and oceanography, magnetic and gravitational anomalies.

8.3   Methodology and Plan for 3D Modeling

A 3D model of an asteroid may be generated directly from a series of images of a spinning object with the axis of rotation perpendicular to the viewer-asteroid line (Fig. 8.1). Boundaries are identifiable as facets rotated through the field of view accompanied by the greatest changes in brightness, slope, curvature, and distance at the facet edge (Fig. 8.2). The undistorted size, shape, orientation and angular relationship of facets could be determined by observing trends in changes in their appearances (foreshortened to undistorted) during each spin. The model complexity

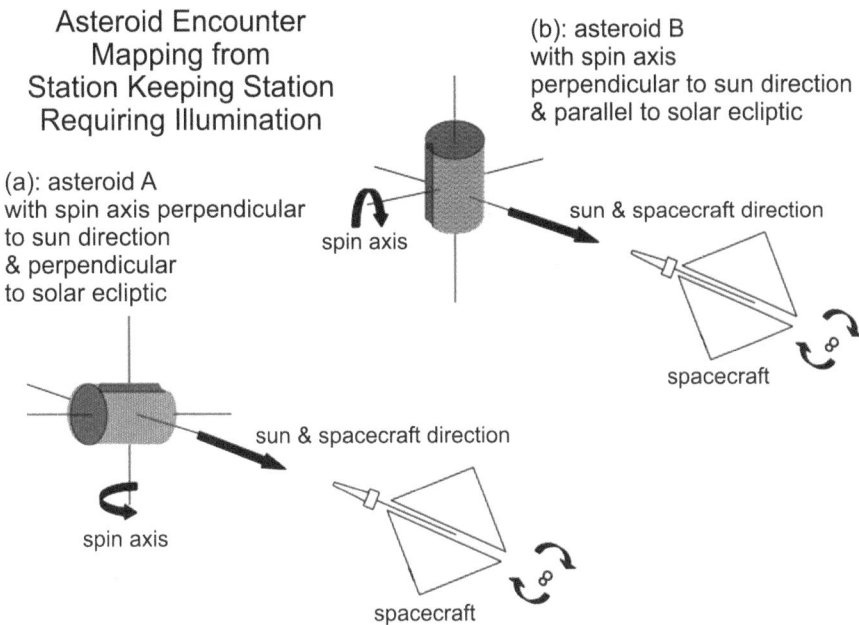

Asteroid Encounter
Mapping from
Station Keeping Station
Requiring Illumination

(b): asteroid B
with spin axis
perpendicular to sun direction
& parallel to solar ecliptic

(a): asteroid A
with spin axis perpendicular
to sun direction
& perpendicular
to solar ecliptic

spin axis

sun & spacecraft direction

spacecraft

sun & spacecraft direction

spin axis

spacecraft

Fig. 8.1 Model of spacecraft-asteroid encounter with orientation needed for mapping drawn by Pamela Clark in 2006

would grow, and branching increases, organically, as the object was approached and the resolution of observations increased.

Ridge lines act as impact event horizons; each impact removed a face and generated an effective standing wave. Ridges form a nested interference pattern of these waves. In principle, the CSNB mapping technique would be well-suited to autonomous operation.

The rotating potato demonstration is also useful in illustrating CSNB map creation from an irregular object (Fig. 8.3). We create a highly faceted object by slicing faces from the potato, as impact events do on an asteroid. Peel remaining between potato facets represents ridges between impacted facets of an asteroid. We observe the potato from a point of view perpendicular to the axis of rotation, progressively increasing resolution of captured images and complexity of captured boundaries, in a manner analogous to asteroid encounter with progressively greater proximity in space. Information is limited to an outline against the dark of space (e.g., the potato against the file cabinet background). A map is created, by connecting maximum elevations along ridges through saddles, using hinges, and minimizing internal distortion. With the remaining potato peel acting as boundaries, we identify and determine the shape and angular relationship of faces by observation during the rotation. In one turn (ABCDE), note movement of natural edges, acting as strings of 'control points', identified and assigned an orientation. A facet maximum size is captured when it is perpendicular to a normal line from the observer. Its orientation, relative to other facets, is determined by its times of appearance and disappearance at the proper angular rotation rate.

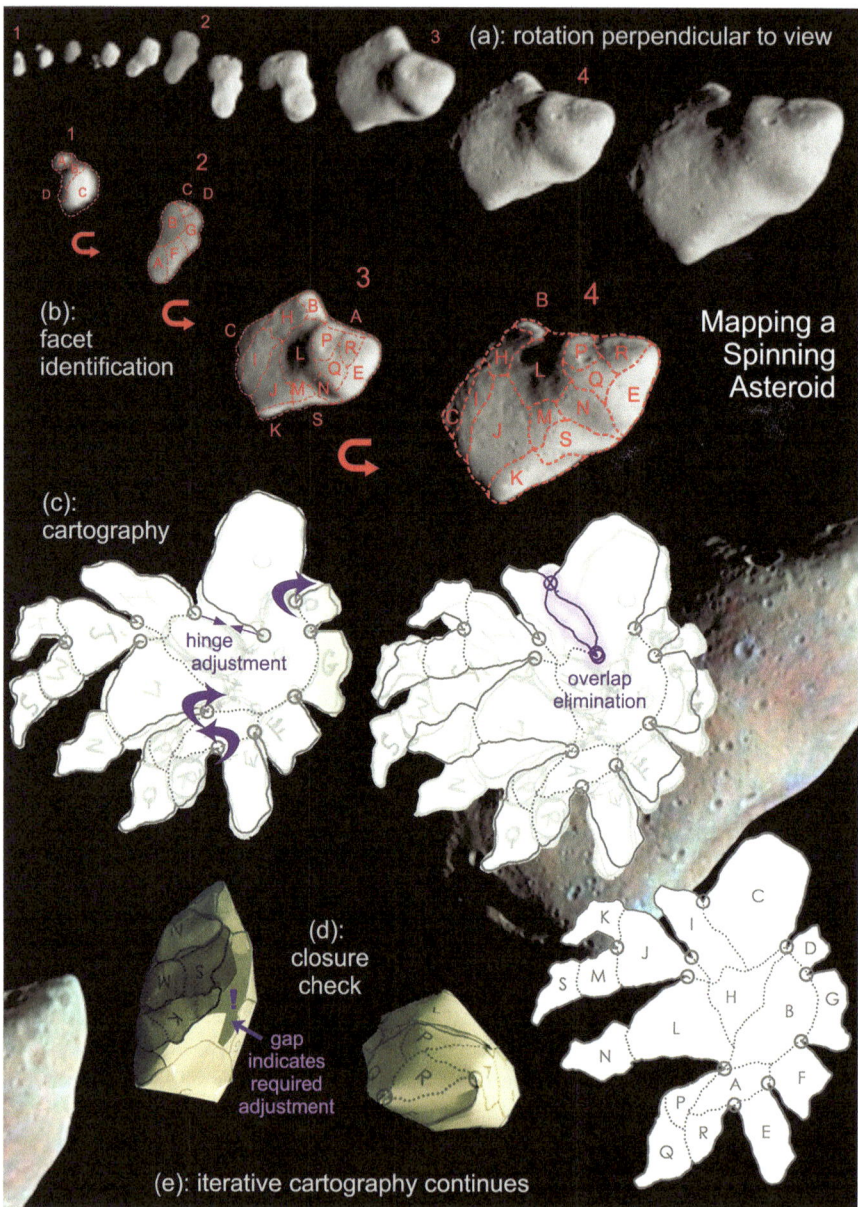

Fig. 8.2 Progressive identification of facets to form a 3D model during spacecraft-asteroid encounter as described in text. (**a**) Courtesy of NASA/JPL, Galileo Gaspra approach sequence; (**b**) and (**c**) Clark (2007)

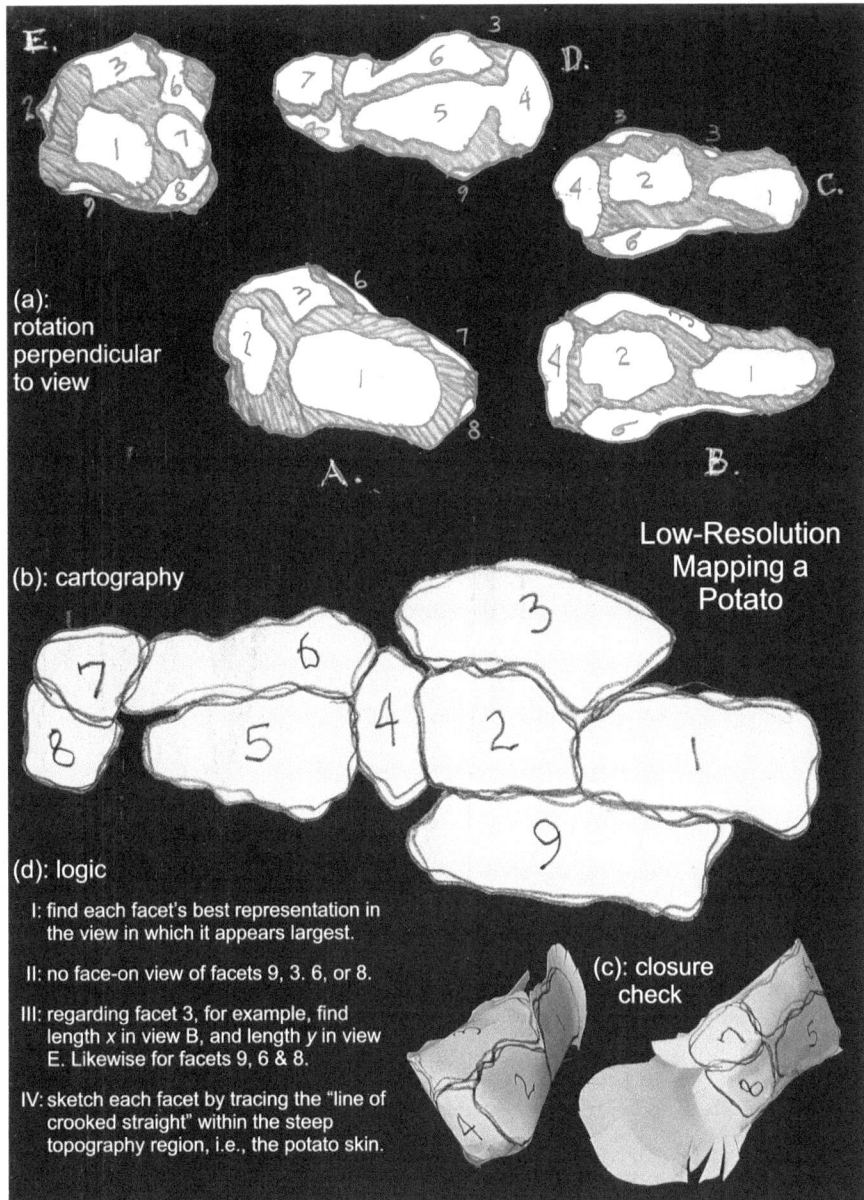

Fig. 8.3 Identifying boundaries and mapping a rotating potato (with peel removed to simulate asteroid facets) from which a 3D model can be created to simulate an analogous asteroid encounter process (Clark and Clark 2006c)

References

Ackerman, J.: New eyes on the oceans. Natl. Geogr. **198**(4), 86–114 (2000)

Aczel, A.D.: The Artist and the Mathematician: The Story of Nicolas Bourbaki, the Genius Mathematician Who Never Existed. High Stakes Publishing, London (2007)

Baltrusaitis, J.: Anamorphoses ou Thaumaturgus opticus, 3rd edn. Flammarion, Paris (1984)

Banerdt, W.A., Vidal, A.: Surface drainage on Mars. Lunar and Planetary Science XXXII, 1488. pdf (2001)

Barbaro, D.: La Pratica della Perspettiva de Monsignor Daniele Barbaro Eletto Patriarca D'Aquileia, Opera Molto Utile a Pittori, a Scultori, et ad Architetti. Camillo and Rutilio Borgominieri, facsimile edition. Bologna (1977–1978), Venice (1569)

Bennett, C.L., Hill, R., Hinshaw, G., Larson, D., Smith, K., Dunkley, J., Gold, B., Halpern, M., Jarosik, N., Kogut, A., Komatsu, E., Limon, M., Meyer, S., Nolta, M., Odegard, N., Page, L., Spergel, D., Tucker, G., Weiland, J., Wollack, E., Wright, E.: Seven-year wilkinson microwave anisotropy probe (WMAP) observations: are there cosmic microwave background anomalies? Astrophys. J. Suppl. **192**, 17–36 (2011)

Berthoud, M.G.: An equal area map projection for irregular objects. Icarus **175**, 382–389 (2005)

Blum, H.: A transformation for extracting new descriptors of shape. In: Wathen-Dunn, W. (ed.) Models for the Perception of Speech and Visual form, Proceedings of a Symposium, Air Force Research Laboratory, pp. 362–380. MIT Press, Cambridge (1967)

Boyer, C.B.: A History of Mathematics, pp. 188–189. Princeton University Press, Princeton (1968). 329, 653, 668–669

Brouwer, L.: Intuitionism and formalism. Dresden, A. (trans.). Bull. Am. Math. Soc. **20**(2), 81–96 (1913)

Cadogan, P.H.: Oldest and largest lunar basin? Nature **250**(5465), 315–316 (1974)

Carr, M.H.: The Surface of Mars. Sky Publishing, Cambridge, MA (2000)

Cartwright, M.L.: Mathematics and thinking mathematically. Am. Math. Mon. **77**(1), 20–28 (1970)

Cheng, A., Izenberg, N., Chapman, C., Zuber, M.: Ponded deposits on Eros. Meteorit. Planet. Sci. **37**, 1095–1105 (2002)

Christensen, A.: The revival of a victorian art: waterlining with a computer. Br. Cartogr. Soc. **36**, 1 (1999)

Clark, C.S., with Jones, LT: Anamorphic Couch, Oils on wallboard, 3 × 8, Spaces & Illusions, High Museum of Art, Atlanta, GA (1977–1983).

Clark, C.S., with Lamb, M.: Robert Frost Made Me Do It!, latex on gatorboard, fir, aluminum, Sunday funnies, peanut butter jar, 6 × 10 × 11, Spaces & Illusions, High Museum of Art, Atlanta, GA (1977–1983).

P.E. Clark and C. Clark, *Constant-Scale Natural Boundary Mapping*
to Reveal Global and Cosmic Processes, SpringerBriefs in Astronomy,
DOI 10.1007/978-1-4614-7762-4, © The Author(s) 2013

Clark, C.S., with Williams: S.C., Fata Morgana, Oil on wallboard and mirror-chrome steel, 7 × 12 × 10, Spaces & Illusions at the High Museum of Art, Atlanta, GA (1977–1983)

Clark, C.S.: World maps with constant-scale natural boundaries and the asteroid Eros. Lunar and Planetary Science XXXIII, 1794.pdf (2002).

Clark, C.S.: Visual calculus or perceptual fribble? World maps with constant-scale natural boundaries: A novel projection method well suited to our era. Advances in Extraterrestrial Mapping, ISPRS WG-IV/7, 34, Lunar and Planetary Institute (2003).

Clark, C.S.: Illustrating concepts in global tectonics with world maps with constant-scale natural boundaries. New Concepts in Global Tectonics Newsletter **31**, 15–19 (2004a).

Clark, C.S.: Mars crustal dichotomy and world maps with constant scale natural boundaries: A creative approach to visualizing subtle points of geodesy. Workshop on Hemispheres Apart, Contribution No. 1213, 4020.pdf, Lunar and Planetary Institute (2004b).

Clark, C.S.: The Martian watershed, geology, and paleohydrology on two world maps with constant-scale natural boundaries. Lunar and Planetary Science XXXVI, 2189.pdf (2005).

Clark, C.S.: World maps with constant-scale natural boundaries: CSNB 2007. Advances in Planetary Mapping 2007, ISPRS WG IV/7 Extraterrestrial Mapping Workshop, 8–9. Lunar and Planetary Institute (2007).

Clark, C.S.: Foldable boundary-based world maps of geologies of Enceladus and Ganymede. Lunar and Planetary Science XLII, 2809.pdf (2011).

Clark, C.S. and Clark, P.E.: Using boundary-based mapping projections to reveal patterns in depositions and erosional features on 433 Eros. Lunar and Planetary Science XXXVII, 1189.pdf (2006a).

Clark, C.S. and Clark, P.E.: Venus on two world maps with constant-scale natural boundaries. Advances in Planetary Mapping 2007, ISPRS WG IV/7 Extraterrestrial Mapping Workshop, 11–12. Lunar and Planetary Institute (2007).

Clark, C.S. and Clark, P.E.: Using boundary-based mapping projections for morphological classification of small bodies. Lunar and Planetary Science XL, 1133.pdf (2009).

Clark, C.S. and Clark, P.E.: Using boundary-based mapping to determine underlying structure for Itokawa and other small bodies. Lunar and Planetary Science XLI, 1264.pdf (2010).

Clark, C.S., Stooke, P.J., Clark, P.E., De Hon, R.A: A more topological planetary cartography: world maps with constant-scale natural boundaries (CSNB). Lunar and Planetary Science XXXVII, 1207.pdf (2006).

Clark, P.E.: Geochemical differentiation of the maria on the early Moon. Lunar and Planetary Science XVI, 137–138 (1985).

Clark, P.E. and Clark, C.S.: Constant-scale natural boundary mapping as tool for characterizing asteroids. Lunar and Planetary Science XXXVI, 1432.pdf (2005).

Clark, P.E. and Clark, C.S.: Using boundary-based maps to illustrate the palimpsest effort of early impacts on lunar surface formation. Lunar and Planetary Science XXXVII, 1153.pdf (2006b).

Clark, P.E. (PI) and Clark, C.S. (Co-I): Revolutionary Morphology-based World Mapping, Proposal, NIAC CP 06-01 Phase 1 (2006c).

Clark P.E., Clark, C.S., Lowman, P.D., Jr.: Applying the constant-scale natural boundary mapping technique. Advances in Planetary Mapping 2007, ISPRS WG IV/7 Extraterrestrial Mapping Workshop, 12–13. Lunar and Planetary Institute (2007).

Clark, P.E. and McFadden, L.A.: New results and implications for lunar crustal iron distribution using sensor data fusion techniques. Geophysical Research Letters: Planets **105**, 4291–4316 (2000).

Clark, P.E. and Rilee, R.L.: Remote Sensing Tools for Exploration. 300 pp. Springer, New York (2010).

Connerney, J.E.P.: A magnetic perspective on the Martian crustal dichotomy. Lunar and Planetary Institute Workshop on Hemispheres Apart, Contribution No. 1213.pdf. Lunar and Planetary Institute (2004)

Copi, C.J., Huterer, D., Schwarz, D., Starkman, G.: On the large-angle anomalies of the microwave sky. Mon. Notices. R. Astron. Soc. **367**, 79–102 (2006). doi:10.1111/j.1365-2966.2005.09980.x

Copi, C.J., Huterer, D., Schwarz, D., Starkman, G.: Large-angle anomalies in the CMB. Adv. Astron. **2010**, 847541 (2010). doi:10.1155/2010/847541. 17 pp

Coxeter, H.S.M.: Regular Polytopes, p. 321. Dover, Mineola (1973)

da Vinci, L.: Two views of the skull, pen and brown ink over black chalk, 188×134 mm. Royal Library, Windsor Castle (1489)

De Hon, R.A.: Hydrogeologic provinces of Mars. Lunar. Planet. Sci. **XXVI**(26), 327–329 (1995)

De Hon, R.A.: Hydrologic provinces of Mars: Physiographic controls on drainage and ponding. In: Cabrol, N., Grin, E. (eds.) Lakes on Mars, pp. 68–89. Elsevier, Amsterdam (2010)

de Vaulezard, I.L.: Perspective Cilindrique et Conique, our Traicte des Apparences Veues par le Moyen des Miroirs Cilindrique et Conique. J. Jacquin, Paris (1630)

Diego, J.M., Cruz, M., Gonzalez-Nuevo, J., Maris, M., Ascasibar, Y., Burigana, C.: WMAP anomalous signal in the ecliptic plane. Mon. Notices. R. Astron. Soc. **402**(2), 1213–1220 (2010)

Dohm, J.M., Kerry, K., Keller, J., Baker, V.R., Boynton, W., Maruyama, S., Anderson, R.C.: Mars geological province designations for the interpretation of GRS data. Lunar and Planetary Science XXXVI, 1567.pdf (2005)

Dohm, J.M., Ferris, J.C., Baker, V., Anderson, R.C., Hare, T.M., Strom, R.C., Barlow, N.G., Tanaka, K.L., Klemaszewski, J.E., Scott, D.H.: Ancient drainage basin of the Tharsis region Mars: potential source for outflow channel systems and putative oceans or paleolakes. J. Geophys. Res. Planet. **106**(E12), 32,943–32,958 (2001)

Dürer, A.: St. Jerome in his study, copperplate engraving, 9.7×7.4 inches. (1514)

Dürer, A.: Underweysung der messing mit dem zirckel und richtscheyt in linien, ebnen, und gantzen corporen, durch Albrecht Dürer zusammengezogen und zu nutz alle kunstliebhabenden mit zu gehorigen Figuren, in truck gebracht/Manual of measurement by compass and straightedge of lines, areas, and solids; assembled with appropriate illustrations by Albrecht Dürer for use by all students of art. Abaris Books, Nuremburg (1525)

Edgett, K.S.: The sedimentary rocks of Sinus Meridiani: Five key observations from data acquired by the Mars Global Surveyor and Mars Odysssey Orbiters. Mars, 1, 5–58, doi:10.1555/mars.2005.0002 (2005).

Edgett, K.S. and Parker, T.J.: Regarding a wet, early Noachian Mars: Geomorphology of western Arabia and northern Sinus Meridiani, Conference on Early Mars: Geologic and Hydrologic Evolution, Physical and Chemical Environments, and the Implications for Life, 3020.pdf (1997)

Fairén, A.G., Dohm, J.M., Baker, V.R., de Pablo, M.A., Ruiz, J., Ferris, J.C., Anderson, R.C.: Episodic flood inundations of the northern plains of Mars. Icarus **165**(1), 53–67 (2003). doi:10.1016/S0019-1035(03)00144-1

Fisher, I. and Miller, O.: World Maps and Globes, New York Essential Books (1944)

Frey, H., Roark, J., Shockey, K., Frey, E., Sakimoto, S.: Ancient lowlands on Mars. Geophys. Res. Lett. **29**, 10 (2002). doi:10.1029/2001GL013832

Goldstein, S.J.: Right cerebellar tumor with mild edema and mass effect. MedPix image no. 26246 (2009)

Greeley, R., and Batson, R.: Planetary Mapping, Cambridge University Press, 312 p. (1990)

Greenberg, J., Jordan, S.: Chuck Close up Close. DK Publishing, New York (1998)

Hausdorff, F.: Bemerkung über den inhalt von punktmengen. Math. Ann. **75**, 428–433 (1914)

Hinshaw, G., Nolta, M., Bennett, C., Bean, R., Dore, O., Greason, M., Halpern, M., Hill, R., Jarosik, N., Kogut, A., Komatsu, E., Limon, M., Odegard, N., Meyer, S., Page, L., Peiris, H., Spergel, D., Tucker, G., Verde, L., Weiland, J., Wollack, E., Wright, E.: Three-year Wilkinson microwave anisotropy probe (WMAP) observations: temperature analysis. Astrophys. J. **170**, 288 (2007)

Hirata, N., Barnouin-Jha, O.S., Honda, C., Nakamura, R., Miyamoto, H., Sasaki, S., Demura, H., Nakamura, A., Michikami, T., Gaskell, R., Saito, J.: A survey of possible impact structures on 25143 Itokawa. Icarus **200**, 486–502 (2008)

Jacobs, F.: A foldable map of Mars's moon phobos. In: Strange Maps: An Atlas of Cartographic Curiosities, p. 202. Viking Studio/Penguin Group, New York (2009)

Keele, K.D.: Leonardo da Vinci the anatomist, in Leonardo da Vinci Anatomical Drawings from the Royal Library Windsor Castle, pp. 10–14, Metropolitan Museum of Art, New York (1983)

Kim, J., Naselsky, P., Hansen, M.: Symmetry and antisymmetry of the CMB anisotropy pattern. Adv. Astron. **2012**, 960509 (2012)

Kovac, J., Leitch, E., Pryke, C., Carlstrom, J., Halverson, N., Holzapfel, W.: Detection of polarization in the cosmic microwave background using DASI. Nature **420**, 772–787 (2002)

Krantz, S.: Conformal mappings. Am. Sci. **87**, 436–445 (1999)

Kreslavsky, M.A., Head, J.W.: New observational evidence of global seismic effects of basin-forming impacts on the moon from lunar reconnaissance orbiter lunar orbiter laser altimeter data. J. Geophys. Res. **117**, E00H24 (2012). doi:10.1029.2011JE003975

Lagerstrom, R., Sun, C., Vallotton, P.: Boundary extraction of linear features using dual paths through gradient profiles. Pattern. Recognit. Lett. **29**, 1753–1757 (2008)

Lakdawalla, E. S.: Phobos arts and crafts. The planetary society blog. http://www.planetary.org/blogs/emily-lakdawalla/2008/1348.html (2008). Accessed 2008

Landau, D., Chase, J., Randolph, T., Timmerman, P., Oh, D.: Electric propulsion system selection process for interplanetary missions, J Spacecraft and Rockets, 48, 3, 467–476 (2011)

Leupold, J.: Theatri Machinarum Generale: Schau-Platz des Grundes mechanischer Wisssenschafften, deutliche Anleitung zur Mechanic oder Bewegungs-Kunst. Published by the author and Gleditschens, J.F., Leipzig. Kinematic models for design digital library. http://ebooks.library.cornell.edu/cgi/t/text/text-idx?c=kmoddl;idno=kmod022 (1724)

Lowman, P.D.: Landsat and apollo: the forgotten legacy. Photogramm. Eng. Remote. Sens. **65**, 1143–1147 (1999)

Lowman, P.D.: Exploring Space, Exploring Earth: New Understanding of the Earth from Space, p. 382. Cambridge University Press, Cambridge (2002)

Lowman, P.D.: Global tectonic activity map. NASA/GSFC. http://denali.gsfc.nasa.gov/dtam/gtam/. http://denali.gsfc.nasa.gov/research/lowman/lowman.html (2013). Accessed 2013

Maizlish, A.: Prominence and orometrics: a study of the measurement of mountains. http://www.peaklist.org/theory/theory.html (2003)

Mantz, A. B., Sullivan R. J., Veverka J.: Downslope regolith movement in craters on Eros. Lunar and Planetary Science XXXIII, 1851.pdf (2002)

Marolois, S.: La Perspective Contenant tant la Theorie que la Practique et Instruction Dondamentale d'Icelle. A la Haye chez Henr. Hondius, Paris (1614)

Marr, D.: Vision: A Computational Investigation into the Human Representation and Processing of Visual Information. W.H. Freeman, San Francisco (1982)

Mather, J., Fixsen, D., Shafer, R., Mosier, C., Wilkinson, D.: Calibrator design for the COBE far infrared absolute spectrophotometer. Astrophys. J. **512**, 511 (1999)

Maxwell, J.C.: On hills and dales. Philos. Mag. **40**(269), 421–427 (1870). doi:10.1080/14786447008640422 (2009)

Maxwell, J.C., with remarks by Professor Forbes: On the descriptions of oval curves and those having a plurality of foci. Proceedings of the Royal Society of Edinburgh **II**, pp. 89–91 (Scientific Papers of JCM 1: 1–3) (1846)

Maxwell, J.C.: Manuscript of trifocal curves. From the original in the University Library, Cambridge Scientific Papers of JCM 1) (1847)

McNamee, J.B., Borderies, N.J., Sjogren, W.L.: Venus global gravity and topography. J. Geophys. Res. Planet. **98**(E5), 9113–9128 (1993). doi:10.1029/93JE00382

Miller, G.A.: The magical number seven, plus or minus two: some limits on our capacity for processing information. Psychol. Rev. **63**(2), 81–97 (1956). doi:10.1037/h0043158

Minor Planet Center: http://www.minorplanetcenter.org/iau/mpc.html (2012). Accessed 2012

Miyamoto, H., Yano, H., Scheeres, D., Abe, S., Barnouin-Jha, O., Cheng, A., Demura, H., Gaskell, R., Hirata, N., Ishiguro, M., Michikami, T., Nakamura, A., Nakamura, R., Saito, J., Sasaki, S.: Regolith migration and sorting on asteroid Itokawa. Science **316**(5827), 1011–1014 (2007)

Morse, M.: Mathematics and the arts. Bull. At. Sci. **XV**(2), 55–59 (1959). Special edition edited by Martyl and by Cyril Stanley Smith

Mutch, T.A., Saunders, R.S.: Geologic development of Mars: review. Space. Sci. Rev. **19**(1), 3–57 (1976)

Nova Rico: Physical/Political Globe. Rand McNally, Florence (1982)

Öner, T. A.: Asteroid models. On C.J. Hamilton Solarviews. http://www.solarviews.com/eng/asteroid.htm (2012). Accessed 2012

Panofsky, E.: Albrecht Durer. Princeton University Press, Princeton (1943)

Parker, T.J., Grant, J.A., Anderson, F.S., Franklin, B.J.: MOLA topographic evidence for a massive Noachian ocean on Mars. Lunar and Planetary Science XXXIII, 2027.pdf (2002)

Rana, S.: Issues and further directions. In: Rana, S. (ed.) Topological Data Structures for Surfaces. Wiley, Chichester (2004). 200 pp

Saito, J., Miyamoto, H., Nakamura, R., Ishiguro, M., Michikami, T., Nakamura, A.M., Demura, H., Sasaki, S., Hirata, N., Honda, C., Yamamoto, A., Yokota, Y., Fuse, T., Yoshida, F., Tholen, D.J., Gaskell, R.W., Hashimoto, T., Kubota, T., Higuchi, Y., Nakamura, T., Smith, P., Hiraoka, K., Honda, T., Kobayashi, S., Furuya, M., Matsumoto, N., Nemoto, E., Yukishita, A., Kitazato, K., Dermawan, B., Sogame, A., Terazono, T., Shinohara, C., Akiyama, H.: Detailed images of asteroid 25143 Itokawa from Hayabusa. Science **312**(5778), 1341–1344 (2006). doi:10.1126/science.1125722

Siddiqi, K., Pizer, S.: Medial Representations: Mathematics, Algorithms, and Applications. Springer, New York (2005)

Sidoli, N., Berggren, J.L.: The Arabic version of Ptolemy's planisphere or flattening the surface of sphere: text, translation, commentary. SCIAMVS **8**, 37–139 (2007)

Smith, D.E., Zuber, M.T., Frey, H.V., Garvin, J.B., Head, J.W., Muhleman, D.O., Pettengill, G.H., Phillips, R.J., Solomon, S.C., Zwally, H.J., Banerdt, W.B., Duxbury, T.C., Golombek, M.P., Lemoine, F.G., Neumann, G.A., Rowlands, D.D., Aharonson, O., Ford, P.G., Ivanov, C.L., Johnson, C.L., McGovern, P.J., Abshire, J.B., Afzal, R.S., Sun, X.: Mars orbiter laser altimeter: experiment summary after the first year of global mapping of Mars. J. Geophys. Res. Planet. **106**(E10), 23689–23722 (2001). doi:10.1029/2000JE001364

Snyder, J.P.: A Comparison of pseudocylindrical map projections: American Cartographer, 4, 1, 59–81 (1977)

Snyder, J.P.: Space Oblique Mercator Projection Mathematical Development. USGS Bulletin 1518 (1981).

Snyder, J.P.: Flattening the Earth: Two Thousand Years of Map Projections. University of Chicago Press, Chicago (1993). ISBN 0-226-76747-7

Snyder, J.P., Voxland, P.M.: An album of map projections. USGS professional paper 1453, US Government Printing Office (1989)

Spilhaus, A.F.: Atlas of the World with Geophysical Boundaries: Showing Oceans, Continents, and Tectonic Plates in their Entirety, Memoir 196. American Philosophical Society, Philadelphia (1991)

Spilhaus, A.F., Snyder, J.P.: World maps with natural boundaries. Cartogr. Geogr. Inf. Sci.–Am. Congr. Surv. Mapp. **18**(4), 246–254 (1991). doi:10.1559/152304091783786709

Stooke, P.J.: Mapping worlds with irregular shapes. Can. Geogr. **42**(1), 61–78 (1998)

Stooke, P.J.: Planetary maps and images, small world atlas 2000. V1.o Multi-SA-Multi-6-Stookemaps-V1.0.NASA PDS. http://www.ssc.uso,ca/geography/space-map/contents.htm (2012). Accessed 2012

Thomas, P., Joseph, J., Carcich, B., Veverka, J., Clark, B.E., Bell, J.F., Byrd, A., Chomko, R., Robinson, M.: Eros: shape, topography, and slope processes. Icarus **155**(1), 18–37 (2002)

Tobler, W.: Map Transformations of Geographic Space, Ph.D. Thesis, U Washington, Seattle, (U. Microfilm #61-4011) (1961)

USGS: Map showing relief and surface markings of the lunar polar regions. I1236A (1981)

USGS: Map showing relief and surface markings on the lunar far side. I1218A (1980)

USGS: Shaded relief map of the lunar near side. I2276B (1992)

van Thiel, P.J.J.: Rembrandt 1669–1969. ASIN B002DOTOAS. Rijksmuseum, Amsterdam (1969)

van Wijk, J.J.: Unfolding the earth: myriahedral projections. Cartogr. J. **45**(1), 32–42 (2008)

Veltman, K.H.: Perspective, anamorphosis, and vision. Marbg. Jahrb. **21**, 93–117 (1986)

Vignola, J.: La Due Regole della Prospettiva di M. Iocomo Barozzi da Vignola con I Comentarij del R.P.M. Egnatio Danti. Facsimiles Edition, Rome (1974), Bologna (1583)

von Helmholtz, H.: Helmholtz's Treatise on Physiological Optics. Southall, J. (trans.). The Opt. Soc. Am., Washington DC (1924)

Weyl, H.: Die Idee der Riemannschen Fläche. B.G. Teubner, Wiesbaden (1997) Leipzig (1913)

Whitaker, E.A.: The lunar Procellarum basin. In Multi-ring Basins, Proceedings of the Lunar and Planetary Science conference, Schultz, P. and Merrill, R., Eds., 12A, 105–111, Pergamon Press (1981)

Whitmore, T.C.: Wallace's Line and Plate Tectonics. Clarendon Press, Oxford (1981). 90 pp

Wilhelms, D.: Procellarum, a giant planetary basin. NASA Technical Memorandum 85127, Reports of the Planetary Geology Program, NASA/OSSA (1982)

Wilhelms, D.: The Geologic History of the Moon. USGS Professional Paper 1348, US Government Printing Office (1987)

Wilhelms, D., Squyres, S.: The Martian hemispheric dichotomy may be due to a giant impact. Nature **309**, 138–140 (1984)

Wollaston, W.: On the apparent direction of eyes in a portrait. Philos. Trans. R. Soc. London **B114**, 247–260 (1824)

Young, D., Harrison, J.F.: Lone pairs and the electrostatic potential of water. Computational Chemistry List, Ltd. http://www.ccl.net/cca/documents/dyoung/water/lp3.shtml (2001). Accessed 2012

Zaidel, D., Hessamian, M.: Asymmetry and symmetry in the beauty of human faces. Symmetry **2**, 136–149 (2010). doi:10.3390/sym2010136

Zoback, M.A.: First and second order patterns of stress in the lithosphere: the World Stress Map Project. Geophys. Res. Lett. **97**, B8 (1992)

Provenances

We are adding this section to provide interesting details on provenances: the sources of data used in creating the drawings and maps you see in the figures. We don't always provide details for portions of figures with captions that clearly describe their sources. Dates represent at least approximately when the material was created.

Fig. 1.1a (From Snyder 1993) Egyptians and Greeks (c200 B.C.)

Fig. 1.1b (From Snyder 1993) Ancient Egyptians (unknown date)

Fig. 1.1c (From Snyder 1993) Gerhardus Mercator: Nova et Aucta Orbis Terrae Descriptio ad Usum Navigantium Emendate Accommodata (New and more complete Representation of the Terrestrial Globe properly adapted for Use in Navigation). Duisburg (1569)

Fig. 1.1d (From Panofsky 1943) Albrecht Dürer: 'Net' of the Cuboctohedron Truncum, fourth book on geometry. Saxon State Library, Dresden (1519)

Fig. 1.2a Created by Chuck Clark with NASA GSFC data (2003)

Fig. 1.2b, c As in Fig. 1.1b, c (2012)

Fig. 1.2d Created by Chuck Clark from method described by Dürer (2003)

Fig. 1.3a C.S. Clark with jim hagan and Richard Noone: Map of Earth centered on the south rotational pole, showing geometrically correct areas, shapes, and adjacencies, with Ocean currents superposed, which folds to a sphere exaggerated to depict Earth's known pear shape. Pin-bar-drafted photographic print and mixed media, 35×47 in. Presented Georgia Geographic Alliance annual meeting, Atlanta (1993)

Fig. 1.4c C.S. Clark: Eros Map 2012. Digital media, 240 megabytes. Unpublished (2012)

Fig. 1.5a, b Topography data from Cheng et al. (2002), patch and pond (Data from Mantz et al. (2002))

Fig. 1.5c From Peter Thomas (2006)

Fig. 1.6a C.S. Clark modified Chart of Inland Basin (p. 240) in Maxwell (1870) "On Hills and Dales".

Fig. 1.6b–e C.S. Clark (Adapted from conference poster (2002))

Fig. 1.7a C.S. Clark: boundary components, Earth study. Unpublished (1999)

Fig. 1.7b C.S. Clark: conformal elbow. Unpublished (2012)

Fig. 1.8 C.S. Clark: Earth, showing Oceans, with watersheds and rivers in true shape, size, and adjacency. Ink on Mylar, 33×43 in. Constant-scale natural boundaries traced from Rand McNally physical/political Globe (© 1982 Rico Florence) (1994)

Fig. 1.9 From NASA/GSFC global tectonic activity Map via Paul Lowman (2000)

Fig. 1.10b C.S. Clark with jim hagan: unpremiated entry, Gondwana circle design competition. San Francisco Botanical Garden, San Francisco (2010)

Fig. 1.11 Foreground C.S. Clark: CSNB human head, anonymous (Lee). Graphite on Mylar, 23×29 in. (2000)

Fig. 1.11 Background C.S. Clark: *Pixelated Paul*. Digital media, 320 megabytes. Unpublished (2012)

Fig. 1.12 From Tayfun Öner, extraordinary model maker (2012)

Fig. 1.13 C.S. Clark coordinated by Scogin Heery: *Spaces & Illusions* at the High Museum of Art, Gudmund Vigtel, director. Funding by National Endowment for the arts. 3,000 ft². Atlanta (1977–1983)

Fig. 1.13a C.S. Clark with L T Jones: *Anamorphic Couch*. Oil on wallboard, 3×8 ft (1978, razed 1983)

Fig. 1.13b C.S. Clark with Susan Clark Williams: *Fata Morgana*. Oil on wallboard, oil on plywood, chrome steel, 7×12×10 ft (1978, razed 1983)

Fig. 1.13c C.S. Clark from a suggestion by Merrill Elam: *Robert Frost Made Me Do It! (Anamorphic Dining Room)*. Latex on Gatorboard and fir, aluminum, Sunday funnies, peanut butter jar, 6×10×11 ft (1978, razed 1983)

Fig. 2.1, **2.2**, **2.3**, and **2.4** René De Hon and C.S. Clark: Mars study. Unpublished (2004)

Fig. 2.5 from NASA/GSFC global tectonic activity map via Paul Lowman (2003)

Fig. 2.6 from NASA MOLA topography data. Created with help of R. De Hon (2004)

Fig. 2.7 from NASA/GSFC global tectonic activity map via Paul Lowman (1998)

Fig. 2.8 C.S. Clark: waterlines study (Adapted from conference poster (2002))

Fig. 2.9 C.S. Clark (Adapted from conference demonstration (2007))

Fig. 2.10 C.S. Clark, suggested by P.E. Clark and Gunther Kletestchka (Adapted from conference poster (2007))

Fig. 4.1 From NOAA MGG data (2006)

Fig. 4.2 From Geological Map of the World at 1:250,000,000, second Edition, Bouysse et al. © CCGM/CGMW 2000 (2007)

Fig. 4.3 From NASA/GSFC global tectonic activity map via Paul Lowman (2006)

Fig. 4.4b C.S. Clark: cross-section of paper joints. www.rightbasicbuilding.com (2011)

Fig. 4.5 C.S. Clark: 'M' and 'N': complementary watershed maps, with uninterrupted oceanographics. Unpremiated entry, American Congress on Surveying and Mapping 27th annual map design competition, Gaithersburg (2001)

Fig. 4.6 C.S. Clark: storm animation, unpublished (Adapted from IRAD proposal (2012))

Fig. 4.7 C.S. Clark: Ocean currents study (Adapted from conference poster (2002))

Fig. 5.1 From USGS publication by Wilhelms (1987) and USGS maps of the Moon (1980, 1981, 1992) (2006)

Fig. 5.2 From NASA LOLA topography data. Conventional maps (**a** and **b**) (From USGS via Trent Hare (2012))

Fig. 5.3 From NASA maps of Magellan data created by USGS Astrogeological Branch (2007)

Fig. 5.4b René De Hon, original caption: hydrogeologic provinces of Mars. *Short dashed lines* mark divides; solid contours are drawn at 5 km intervals; dashed contours of intermediate intervals show highest and lowest elevations of closure; *dot-and-dash lines* are drainage networks or channels; *heavy solid lines* with arrow mark prominent surface troughs (De Hon 1995)

Fig. 5.4d Alberto Fairén and coauthors, original caption: topographic shaded relief map of the northern hemisphere of Mars constructed from Mars Orbiter Laser Altimeter (MOLA) data showing major geographic features of the northern hemisphere, including three major basins (Borealis basin = Vastitas Borealis, Utopia basin = Utopia Planitia, Isidis basin = Isidis Planitia). Also shown are Shoreline 1 (*black line*), Contact 1 in Arabia Terra (*dashed-black line*) and Shoreline 2 (*dark blue line*), which are based on Edgett and Parker (1997), Carr (2002), Parker et al. (1987, 1989, 2002), Clifford and Parker (2001), and Head et al. (1999); paleolakes (*light blue lines*), based on Scott et al. (1995); and Stage information (numbers) that reflects the geologic mapping of Tanaka et al. (2002) and correlative with Stage information of Dohm et al. (2001b, c) and Anderson et al. (2001). Polar Stereographic projection; scale varies with latitude; modified from Tanaka et al. (2002). (Fairén et al. 2003)

Fig. 5.5 René De Hon and C.S. Clark: Mars study. Unpublished (2004)

Fig. 6.1a NASA/JPL, original caption: model of Asteroid 4769 Castalia. It was the first three-dimensional model of an asteroid ever produced. The picture shows 16 different views of a three-dimensional model of Castalia, which is 1.8 km across at its widest. The model was created by Scott Hudson (Washington State University) and Steve Ostro (JPL) (From data taken at Arecibo Observatory in 1989. Photo no. P43041A.)).

Fig. 6.2 Topography data from Cheng et al. (2002), pond and patches (Data from Mantz et al. (2002))

Fig. 6.3a From Phil Stooke Eros photomosaic (2011)

Fig. 6.3b From Marc Berthoud Eros map projection (2012)

Fig. 6.4 Topography data from Tayfun Öner (2008)

Fig. 6.5 Topography data from Tayfun Öner (2008)

Fig. 6.6 From Deimos and Phobos photomosaics by Phil Stooke and coworkers. orthographic views and topography (Data adapted from Tayfun Öner (2008))

Fig. 6.7 From Eros and Ida photomosaics by Phil Stooke and coworkers. Orthographic views and topography (Data adapted from Tayfun Öner (2008))

Fig. 6.8b Makoto Yoshikawa: analysis of X-ray data, Exploring the asteroid Itokawa. Original caption: reflectance ratio in 1.57 um infrared. *Dark blue* shows reflectance ratio of 0.126 while *red shows* 0.145 (2008) http://www.isas.jaxa.jp/e/forefront/2008/yoshikawa/02.shtml

Fig. 6.8c Contributed by Jun Saito and coworkers on the Hayabusa Project Team (2013)

Fig. 6.9 From photomosaic by Phil Stooke (2010)

Fig. 6.10a From Broken Hill #471 cast specimen of Glen Doran (2007)

Fig. 6.10b, c From Susan Clark Williams collection (2007)

Fig. 6.11 *Top right* Leonardo da Vinci: detail from *Two Views of the Skull. Pen and brown ink* (two shades) over *black chalk*, 188×134 mm. http://www.drawingso-fleonardo.com (1489)

Fig. 6.11 *Top right* Albrecht Dürer: detail from *St. Jerome in His Study*. Copperplate engraving, 9.7×7.4 in. (1514)

Fig. 6.11 *Top right* Dr. Steven J. Goldstein, University of Kentucky College of Medicine: right cerebellar tumor with mild edema and mass effect. MedPix® is an online medical image database of peer-reviewed images, patient profiles, and disease topics. MedPix images meet HIPAA criteria with regard to patient confidentiality. MedPix image no. 26246 (2009)

Fig. 6.11 *Top left* C.S. Clark: monkey cranium Map showing internal surfaces. Watercolor, 19×25 in. presented in a lecture, Clark Atlanta University, Center for Theoretical Structures of Physical Systems, Alfred Z. Msezane, director (1998)

Fig. 6.11 *Left* C.S. Clark: *Boo Too*. Watercolor, 25×22 in. (1998)

Fig. 6.11 *Bottom left water molecule* David Young and James Harrison, original caption: the 0.0327 e/Å3 isosurface with an inscribed sphere (*green*) centered at oxygen. The radius of the inscribed sphere is just large enough to break through the isosurface (2001)

Fig. 6.11 *Bottom middle water molecule* David Young and James Harrison, original caption: two views of the electrostatic potential of the water molecule: the electrostatic potential is contoured both in the molecular plane and perpendicular to the molecular plane bisecting the HOH angle. The innermost contour corresponds to 0.142 e/Å3 and the subsequent contours decrease uniformly by 0.00570 e/Å3 (2001)

Fig. 6.11 *Bottom left Earth* NASA 3D Earth (2012)

Fig. 6.11 *Bottom right* CSNB maps (From Chuck Clark of water molecule (2012) and Earth (2001) (See Fig. 4.5))

Fig. 7.1a COBE http://lambda.gsfc.nasa.gov/product/cobe/dmr_image.cfm. The image shows the reduced map (i.e., both the dipole and Galactic emission subtracted). The cosmic microwave background fluctuations are extremely faint, only one part in 100,000 compared to the 2.73° Kelvin average temperature of the radiation field. The cosmic microwave background radiation is a remnant of the Big Bang and the fluctuations are the imprint of density contrast in the early universe. The density ripples are believed to have given rise to the structures that populate the universe today: clusters of galaxies and vast regions devoid of galaxies (2012)

Fig. 7.1b Seven year microwave sky, NASA/WMAP science team http://wmap.gsfc.nasa.gov/media/101080/index.html: the detailed, all-sky picture of the infant universe created from 7 years of WMAP data. The image reveals 13.7 billion year old temperature fluctuations (shown as color differences) that correspond to the seeds that grew to become the galaxies. The signal from our galaxy was subtracted using the multi-frequency data. This image shows a temperature range of ±200 microKelvin (2011)

Fig. 7.1c Craig Copi and coauthors, original caption: Fig.2. The ecliptic plane and the dashed line is the supergalactic plane. The directions of the equinoxes (EQX), dipole due to our motion through the Universe, north and south ecliptic poles (NEP and SEP) and north and south supergalactic poles (NSGP and SSGP) are shown. The multipole vectors are the solid magenta (medium graying ray scale version) symbols for each map, ILC (*circles*), TOH (*diamonds*), and LILC (*squares*). The open symbols of the same shapes are for the normal vectors for each map. The *dotted lines* are the great *circles* connecting each pair of multipole vectors for this map. The *light gray stars* are particular sums of the multipole vectors which are very close to the temperature minima and maxima of the multipole. The *solid black star* shows the direction of the vector that appears in the trace of the octopole, T3 (29), of the TOH map. The *solid magenta* (again *medium gray* in the *gray scale* version) star is the direction to the maximum angular momentum dispersion for the octopole, again for the TOH map. (Copi et al. 2006)

Figs. 7.2 and **7.3** P.E. Clark and C.S. Clark: cosmic microwave background study. Unpublished (2012–2013)

Fig. 8.1 From P.E. Clark, unpublished (2006)

Fig. 8.2a Gaspra Approach Sequence, PIA00079, original caption: this montage of 11 images taken by the Galileo spacecraft as it flew by the asteroid Gaspra on October 29, 1991, shows Gaspra growing progressively larger in the field of view of Galileo's solid-state imaging camera as the spacecraft approached the asteroid. Sunlight is coming from the right. Gaspra is roughly 17 km (10 miles) long, 10 km (6 miles) wide. The earliest view (*upper left*) was taken 5 3/4 h before closest approach when the spacecraft was 164,000 km (102,000 miles) from Gaspra, the last (*lower right*) at a range of 16,000 km (10,000 miles), 30 min before closest approach. Gaspra spins once in roughly 7 h, so these images capture almost one full rotation of the asteroid. Gaspra spins counterclockwise; its north pole is to the upper left, and the "nose" which points upward in the first image, is seen rotating back into shadow, emerging at lower left, and rotating to upper right. Several craters are visible on the newly seen sides of Gaspra, but none approaches the scale of the asteroid's radius. Evidently, Gaspra lacks the large craters common on the surfaces of many planetary satellites, consistent with Gaspra's comparatively recent origin from the collisional breakup of a larger body. (NASA 1996)

Fig. 8.2b–e C.S. Clark with P.E. Clark: Gaspra mapping (From workshop poster (2007))

Fig. 8.3 P.E. Clark and C.S. Clark: asteroid simulation (Potato). NIAC proposal (2004)

Acronyms

ACSM	American Congress of Surveying and Mapping
CDM	Cold Dark Matter
COBE	COsmic Background Explorer
CSNB	Constant-Scale Natural Boundary (mapping, modeling, method, methodology, technique, tool, approach)
CUA/IACS	Catholic University of America Institute for Astrophysics and Computational Science
CTA	Circular Thin Area (candidate very large impact basins on Mars derived from topography and from crustal thickness data)
GSFC	Goddard Space Flight Center (NASA)
HIPAA	Health Information Portability and Accountability Act
HIRISE	High Resolution Imaging Science Experiment
HEAO	High Energy Astronomy Observatory
IR	Infra Red
JAXA	Japanese Aerospace Exploration Agency
JPL	Jet Propulsion Laboratory (NASA)
LAMBDA	Legacy Archive for Microwave Background Data
LOLA	Lunar Orbiter Laser Altimeter
KREEP	Potassium (atomic symbol K) Rare Earth Elements and Phosporous (a geochemical component of some lunar impact breccia and basaltic rocks)
MESSENGER	MErcury Surface, Space ENvironment, GEochemistry, and Ranging (a robotic NASA spacecraft orbiting the planet Mercury)
MGG	Marine Geology & Geophysics (NOAA)
MRI	Magnetic Resonance Image
MOLA	Mars Orbiter Laser Altimeter
NASA	National Aeronautics and Space Administration (USA)
NCGT	New Concepts in Global Tectonics
NIH	National Institutes of Health
NIRS	National Institute of Radiological Sciences (Japan)

P.E. Clark and C. Clark, *Constant-Scale Natural Boundary Mapping to Reveal Global and Cosmic Processes*, SpringerBriefs in Astronomy, DOI 10.1007/978-1-4614-7762-4, © The Author(s) 2013

NOAA	National Oceanic and Atmospheric Administration (USA)
NOE	Near Earth Objects
OSSA	Office of Space Science and Applications (NASA)
QCD	Quasi-Circular Depression (candidate impact basins on Mars derived from topography and from crustal thickness data)
USGS	United States Geological Survey
WMAP	Wilkinson Microwave Anisotropy Probe

About the Authors

Pamela E. Clark, Ph.D. grew up in New England. Inspired by President John Kennedy, she decided as a child to explore outer space. She thought, "If they can put a man on the moon, they can put a woman (me) on Mars!" She obtained her B.A. from St. Joseph College. There, she had many opportunities to participate in laboratory research with Sr. Chlorophyll (Dr. Claire Markham) and Sr. Moon Rock (Dr. Mary Ellen Murphy) as well as to coordinate an NSF interdisciplinary undergraduate field research project. While obtaining her Ph.D. in planetary geochemistry from the University of Maryland, she worked at NASA/GSFC outside of Washington, DC and the Astrogeology Branch of the USGS in Flagstaff, Arizona, simulating, analyzing, correlating, and interpreting lunar x-ray spectra. She was a member of the group, led by Isidore Adler and Jack Trombka, that pioneered the use of orbital x-ray and gamma-ray spectrometers to determine the composition of planetary surfaces. She participated in the Flagstaff Lunar Data Consortium, the first attempt to create a common format database for all the remote sensing data from a planetary body. After completing her Ph.D., she joined the technical staff at NASA/JPL, worked with the Goldstone Solar System Radar group, and expanded her remote sensing background to include radar, thermal, and near infrared studies of planetary surfaces with particular emphasis on the study of Mercury's surface. Dr. Clark organized a briefing team to promote a mission to Mercury, and for a while edited the Mercury Messenger newsletter. Springer published the first editions of her book "Dynamic Planet: Mercury in the Context of Its Environment" and "Remote Sensing Tools for Exploration". She eventually returned to Goddard to work with the XGRS team on the NEAR mission to asteroid Eros. Dr. Clark is the science lead in a group initiated by Steve Curtis to develop new paradigms for the design of space missions and vehicles. She is currently involved in developing and evaluating surface science scenarios, tools, technologies, and architectures, and for space missions to extreme environments, with particular emphasis on the Moon and Mars. Dr. Clark has done several stints in academia, including Murray State University in Kentucky, Albright College in Reading, Pennsylvania, and Catholic University of America in Washington, DC. She has developed courses in analytical

P.E. Clark and C. Clark, *Constant-Scale Natural Boundary Mapping to Reveal Global and Cosmic Processes*, SpringerBriefs in Astronomy, DOI 10.1007/978-1-4614-7762-4, © The Author(s) 2013

and environmental chemistry, geochemistry, physical geology, mineralogy, optics, planetary astronomy, remote sensing, and physics. Her goals include exploring under every rock to increase the sense of wonder about the solar system.

Chuck Clark, architect: In Massachusetts under the informal tutelage of Robert Frost, Chuck Clark received an unusual childhood education in architecture. Frost, facing a trip to South America and frustrated by conventional maps' "stretchy edges," charged Clark to "Make me a map a *heron* can use . . . to get to Brazil!" At first, Clark succeeded only in tearing his father's road maps, but, undaunted, at age ten he began drafting, and by sixteen was a cartographic technician with the US Army Corps of Engineers in Jacksonville, Florida. Crestfallen to learn from his supervisor, Cleve Powell, that "nobody *draws* world maps anymore. That's all done numerically," he set aside his boyhood task and soon enrolled at Georgia Tech, where he showed promise in perspective drafting, freehand sketching, and partial differential equations. After graduation, he migrated to Mbandaka, Zaire, to help Millard Fuller and others found Habitat for Humanity. Later, relying on Frost's principles – learn to steer water, and to make the part speak for the whole – Clark designed and built museum exhibits that garnered national praise for their "narrative architecture," and inspired Fred Rogers to add a stoplight to the set of his television show, Mr. Roger's Neighborhood. In 1990, Clark was challenged to illustrate certain global symmetries and asymmetries in Earth's land-water distribution that colleague jim hagan had compared to Egypt's Great Pyramid. (Clark's response is within, at Fig. 1.3a.) He next made a map showing watersheds, which he discussed with Athelstan Spilhaus, maker of "world maps with natural boundaries." Clark realized his odd approach was not only distinct from 450 years of tradition, it also used ideas of another childhood mentor, Marston Morse, and, moreover, like his museum exhibits, cloaked form with content. Talks with NOAA's Dave McAdoo and NASA's Paul Lowman led to maps of asteroids and acquaintance with the planetary science community. Conference abstracts and posters over the last decade, and especially the concept-extending insights of coauthor P.E. Clark, have explored the possibilities of global maps with constant-scale natural boundaries. (Clark and Clark share New England roots, and, although they have not done the genealogy, may be distant relatives.) When not mapping, Chuck Clark may be found in Atlanta, Georgia, designing and building pole-type structures, enjoying rambunctious grandchildren, and completing a memoir of his encounters with Frost and Morse.